D0863105

Nothing On
but the Radio

Nothing On but the Radio

*A Look Back
at Radio in
Canada and
How It Changed
the World*

Gil Murray

A HOUNSLOW BOOK
A MEMBER OF THE DUNDURN GROUP
TORONTO · OXFORD

Publisher: Anthony Hawke
Copy-Editor: Michael Hodge
Design: Jennifer Scott
Printer: Transcontinental

National Library of Canada Cataloguing in Publication Data

Murray, Gil, 1925-
    Nothing on but the radio : a look back at radio in Canada and how it changed the world/
    Gil Murray.

ISBN 1-55002-479-5

1. Radio broadcasting--Canada--History. I. Title.

HE8699.C2M87 2003   791.44'0971   C2003-904044-5

1   2   3   4   5   07   06   05   04   03

We acknowledge the support of the **Canada Council for the Arts** and the **Ontario Arts Council** for our publishing program. We also acknowledge the financial support of the **Government of Canada** through the **Book Publishing Industry Development Program** and **The Association for the Export of Canadian Books**, and the **Government of Ontario** through the **Ontario Book Publishers Tax Credit** program, and the **Ontario Media Development Corporation's Ontario Book Initiative**.

Care has been taken to trace the ownership of copyright material used in this book. The author and the publisher welcome any information enabling them to rectify any references or credit in subsequent editions.

                             J. Kirk Howard, President

Printed and bound in Canada.✆
Printed on recycled paper.
www.dundurn.com

Dundurn Press
8 Market Street
Suite 200
Toronto, Ontario, Canada
M5E 1M6

Dundurn Press
73 Lime Walk
Headington, Oxford,
England
OX3 7AD

Dundurn Press
2250 Military Road
Tonawanda NY
U.S.A. 14150

Nothing On but the Radio[1]

table of contents

prologue

From the days of the ancient Greeks, and from the dim times before them, humans have envied the air, envied the creatures who rode upon it, and imagined booming voices like the mythical orator Stentor's, carrying sound on the breeze for many leagues. Some played with echoes of their voices in the mountains, modulating the air by lung pressure. Foiled dreams of flying high on air currents like the Athenian Greek Daedalus and his son Icarus, whose feather-and-wax wings fell apart in the heat of the sun, seemed to condemn humans to walk and talk forever upon only the earth.

Not until the seventeenth century did humans learn to ride the air. First, balloons lifted the daring to great heights. Centuries later, a heavier-than-air plane sailed for a hundred and fifty feet a foot or two above the earth. Its creators reworked the aircraft until it flew across the sky with the grace of a bird. But flying high bodily wasn't enough. Soon human voices began to travel, riding not air currents but the invisible airwaves of radio, striking fear into the hearts of the superstitious, who thought they were hearing voices from Hell.

Belying that fear, a few years later the ardent voices of religious evangelists, hardly from Hell and the Devil, clove the airwaves with messages they declared came from God. The Devil would indeed emerge on the airwaves later. He would be called Television.

The booming far-flung voice of Stentor was eclipsed. Or, perhaps, his ancient prophecy was fulfilled. With the passing of decades voices, music, and a myriad of sounds rode into every nook and cranny on earth, and into space, on the crest of the radio signal. Now it was not the voice of the gods. It was the stentorian voice of the disc jockey, the newscaster, the commentator, the sportscaster, the spacecraft launcher, the gourmet chef . . . but most ubiquitous of all, the commercial message.

In the following pages of lowly print, still a universal medium despite the airwaves, we shall try to bring some airborne stories to earth, to immortalize the mesmerizing ways of the beings — hardly gods — who created the sounds and invoked the mythology of radio itself.

Don't go away.

sparks from on high

The air was blue in studio two. *Somebody*'d had the audacity to chop Canada's most flamboyant newscaster off the air and put on an American crisis report without even the courtesy of a warning. The outrage was unimaginable; it was *outrageous*; it was beyond even that, it was kowtowing to the Americans; it was telling all Canadians that they were second-rate to the Yanks; it was — *x#@#**!x!&***!, to say the least. The whole world would hear about it!

Nobody, but nobody should have the audacity to cut *Gordon Sinclair* off the air in mid-sentence in favour of an American news bulletin when he was in full flight on his ten to six "News and

Commentary" on CFRB, Toronto, Canada's biggest and most successful private radio station. Thus spoke Gordon Sinclair, at any rate. Gord Sinclair cherished having the last word.

So, in the little but powerful world of CFRB's studio two, he was still erupting after the bulletin was over. And when the control room flipped a switch and put him back on the air again in mid-sentence, this time in mid-swearword, blue language flowed out to four hundred thousand listening Ontario households. Swiftly, the news also flowed out to radio stations and newspapers across North America of how a Canadian broadcaster had cursed the institution of U.S. news for reporting on a crisis that might have set off World War III. Telephones rang.

The occasion was the peak of the Cuban Missile Crisis of 1962. Gordon was in the midst of giving a passionate opinion on one of his pet Canadian news sensations. At about mid-point, a signal came over the wire from CBS headquarters in New York, via a magic black box in the CFRB newsroom, that there was a new super-crisis in the ongoing standoff between U.S. President Kennedy and Soviet Union Chairman Khrushchev over the K's insistence on supplying nuclear missiles to Cuba. This posed a serious threat to the United States. The U.S. media viewed it as a possible start to World War III. Gord Sinclair begged to differ — no, he *scorned* the very idea, and had previously said so on-air several times.

Accordingly, on a day when CBS news bulletins were pelting CFRB's newsroom by direct line through the black box several times an hour, and standing RB newsroom orders said they had to go on the air instantly, the inevitable happened. Predictably, one popped up just when Gord Sinclair was on the air. The newsman on duty in the newsroom had his orders: if the black box began to click, he had ten seconds to push the button that put the CBS newsman directly on CFRB's air, regardless of who was on it at the time. The newsman had no choice but to obey. I know, because I was the newsman.

Mine was, therefore, the finger that pushed the button that put the famous Gordon Sinclair off the air and an anonymous American newsman on. Later we'll talk about how that could possibly have happened and what its repercussions were.

But it is a perfect illustration of the power of radio at its peak of public attention — before television moved in and grabbed the ball, then later dropped it. When a potent broadcaster like Gordon Sinclair spoke, hundreds of thousands listening either cheered or booed. Whichever, the voice of radio had, by the 1960s, become the voice of the masses.

What started as a Tinkertoy, an assemblage of gadgets of a kind never before available to the common eccentric, would mushroom into the world's first technological movement based on a non-mechanical physical force — the radio wave. Around it would accumulate an entirely new kind of industry, one based on illusion and persuasion, faith and imagination. Mass audiences in the millions would be created where no such concept had ever existed, audiences who would share the same experiences in the same instant, without ever laying eyes on one another.

It would rely on a listener's imagination, on willingness to believe that the sounds coming from the earphones or the goose-necked horn or the cone-shaped loudspeaker hanging on the wall were from a world more glamorous than one's own. The voices, it was thought, were of people handsome or beautiful, dashing or comical, totally believable and blessed with god-like authority. The music was from some heavenly place, or from ethereal ballrooms found only in the most romantic locales in the universe. Radio came out of the world's first intellectual source borne on an electronic broom that would sweep all minds.

Normally, an author remains invisible in the pages of a book, but in the case of *Nothing On but the Radio*, I, the author, am a small part of *some* of the story and unfortunately must edge out into the light from time to time to tell it as I knew it from the inside. Since radio people ply their trade in person, their stories are best expressed in person, where still possible. However, this account is not an autobiography.

Let us set the scene. While the 1960s may have been when radio's star was at its height, the onset of this magical medium in the 1920s brought disembodied human voices from far away into private homes for the first time. The phenomenon of the airwaves' sounds themselves held the listener in awe. As radio became commercialized, announcers began to speak as "high-class" orators, setting a formal tone for radio for decades to come. On-air talk was short, with the lofty grandiloquence of a superior being. The announcer usually wasn't what he sounded like. Radio people did not yet have The Power. Eventually the god-like announcer was brought down to earth and became a friendly, if overly jocular "DJ," especially in the delivery of commercial "messages."

There was nothing high-class or jovial about radio's earliest days. Amateur home-radio enthusiasts built their own broadcasting equipment that couldn't reach much beyond the block where the house with its radio transmitter was located. Viable stations, however, were soon to blossom all over North America.

Radio content, at first limited to just recorded music and brief announcements, by the 1930s became full-blown programming. Organized U.S. comedy shows such as "Amos 'n Andy" and "Eb 'n Zeb" were as much at home in Canada as in the States. They were overtaken by the even more popular network programs fronted by Bing Crosby, Bob Hope, Edgar Bergen and Charlie McCarthy, Fibber McGee and Molly, Burns and Allen, Eddie Cantor, Jack Benny, Fred Allen, and others. Drawing huge weekly audiences, these U.S. network half-hour shows made even more of an impact than the movies. Laughter is a binding universal force, and during the "dirty thirties" North America needed plenty of laughter.

A North American institution was created when the first "soap operas" rode into Canada on U.S. airways, tying up housewives for hours every afternoon. Most of these made the leap from radio into television later, still tagged with the "soap" identity, even though soap manufacturers, the original radio sponsors, didn't always come with them. "Soaps" they remained. Soap was even at the bottom of radio's bid to become a medium of serious drama when "Lux Radio Theater" hit the radio scene, featuring top Hollywood stars in live hour-long plays, with one of the lords of the movie business, Cecil B. DeMille, as host and introducer. This was not a soap opera, and it didn't make it very far, if at all, into the world of television.

In fact, radio drama was probably more successful in Canada, in Canadian Broadcasting Corporation radio productions like the "Stage" programs. Perhaps our long winter nights called out our imaginations more clearly than did California's enchanted sultry evenings. But even here, live drama transferred to Canadian television didn't survive for long.

In the 1920s, before I, as a five or six year old first became aware of radio around about 1930, dozens of newborn commercial stations across Canada and hundreds in the U.S. had already begun their march upon the public consciousness.

Some believed radio would transform the world — into what, nobody knew. The newspapers of Canada did know one thing: Don't ever mention radio. The fascination for the sounds from the ether rapidly became so great that the papers' owners believed radio was out to kill off their publications.

That is, until certain newspaper tycoons started or bought up radio stations of their own. Though some had second thoughts — like Joseph Atkinson of the *Toronto Daily Star* — and got out of radio broadcasting in the early 1930s, the day came when almost every big newspaper in the country owned one or more stations. After a decade of scattered development of radio stations, and their evolution from being tinpot noise sources to organized entertainment programming with wider geographical coverage, Canadian radio almost came of age.

On July 1, 1927, the Diamond Jubilee of Canada's 1867 Confederation, a loose network of radio stations carried the celebrations in Ottawa across most of the country, to the delight of the population. That was just what was needed. Radio was at last officially recognized as an important public communication force. The federal government thought it wise to set up a national body to keep this new force under control. So in 1930, under Sir John Aird, the Canadian Broadcasting Commission was founded, later to metamorphose into the Canadian Broadcasting Corporation (CBC). Sir John was the one who'd said he *had* owned a radio but he'd "thrown the damn' thing out."

By the late 1930s, the CBC and private stations were awash in programs from the U.S. Public policy makers in Ottawa saw this as bad for the country, but the popularity of the U.S. comedians and other performers kept them at bay from forcing some kind of federal law to block the broadcasts from the south. The outbreak of World War II did the trick. The CBC launched full-scale coverage of the war, sending correspondents overseas to report back. It fit in perfectly with the patriotism brought on by wartime resolve to win. The major daily CBC program became the news at ten p.m., featuring the bass voice of Lorne Greene.

In the small city where I grew up one could listen to the entire CBC ten p.m. newscast while walking down the street on the way home from the early movie at a downtown theatre. On a warm night, all doors and windows lay open, letting out the Greene "Voice of Doom" for every pedestrian to hear. Almost every radio on the street was tuned to CBL Toronto, the chief CBC station. Greene enunciated a grand flow of information every night about the day's war happenings. On arriving home, I felt I'd caught up on all the day's daily war devastation while simply walking home.

The war established radio, especially the CBC, as a household institution. Everyone had a radio, even carrying small portables to the backyard and the beach. Keeping track of the war put radio in the public forefront, giving the newspapers a run for their money. The CBC network became like a latter-day Canadian National Railway, tying the country together as the CNR had in the late nineteenth and early twentieth centuries, introducing East and West directly to each other for the first time with continent-spanning vocal power. By comparison, Stentor would have been a mere yodeller.

But it wouldn't be until the 1950s that stations launched their "chunes-and-chatter" format that would evolve literally into verbal home invasion and go on and on frantically into the next millennium and probably far beyond that. In the 1920s and 30s, people started up radio stations on a shoestring with the same feverish high tech stock market bustle repeated sixty to seventy years later. The potential for making money was great and the new proprietors launched their enterprises with expectations much like those of new automobile dealerships. But radio was changing the world in a far different way from the motorcar. It changed the people.

we've got the way
— now what do we say?

For the early radio pioneers, the answer to that question was simple: WE TALK. At first that was practical, because radio stations didn't stay on the air for long each day, and what the first voices had to say was hardly popular entertainment. In the pre-programming days, practical people like engineers and technicians were really only interested in testing the system and muttering a few bits of wisdom over the air to new radio aficionados, such as tips on signal strength and adjustments to receivers. Radio was still a Tinkertoy for hobbyists. The ambitious amateurs' rudimentary transmitters were tiny — hardly bigger than a breadbox.

Others, ordinary listeners-in, were eagerly fascinated by the very phenomenon itself of hearing sounds in their own homes coming from a distance. Any sound emerging from the crystal set's earphones or even early loudspeakers would amaze, somewhat like the effect that Stentor's booming voice might have had on folks away down the valley.

Programming hadn't arrived yet. But it didn't take long after 1920 for the more loquacious to detect an opportunity for boosting the sound of their own voices, letting them talk to far more people than they might reach otherwise on a soapbox. "Announcers" and "commercials" were born. As audiences matured, talk was fine, but only up to a point, considering the tinny, harsh sounds and the waves of static. Now they needed something to soften the often-indecipherable noises.

The answer was music. But how could they get records — live musicians being so costly — on the air? Play them on a phonograph through a microphone? Not practical. The microphone had to be put right inside the phonograph's horn. Other studio noise got through, too. As at the roots of all great advances, somebody thought there must be a better way. It was indeed coming.

The sounds of music were trapped there in the record grooves, but the thin reproduction produced by the early phonograph pickup heads wasn't good enough for radio broadcasting. The advent of an *electric* pickup head in 1925 enabled music to be fed directly into the broadcasting system from the record, solving the problem of external studio noise. Now the records could roll, feeding music straight onto the airwaves with improved sound. At last there was less talk, more music. Radio became the source of music for Saturday night house parties. The big U.S. dance bands' live network broadcasts from New York, Chicago, and other cities played the latest — emulated by Canadian bands wherever there were professional musicians to perform. Where there were not, stations piped free recorded dance tunes into thousands of homes for Saturday night parties.

The electric pickup head came to the phonograph after years of Bell laboratory research. In a way it was the reverse of the recording head, reproducing the original sounds more faithfully than did the old vibrating diaphragm. In recording, the original master disc was made by cutting bumps into the record grooves with a stylus connected to a recording head with a tiny bit of crystal made from rochelle salt (sodium potassium tartrate) imbedded in it.

Sounds from a live source, such as an orchestra or a singer, were played or sung into a very large horn in an ordinary open room, not an engineered studio. Vibrations passed along a wire produced weak electrical impulses in the crystal. These were then electrically amplified and fed as vibrations to the stylus, which, in the standard way, cut a track into a continuous groove in the master record surface. Wider ranges of high and low sound frequencies resulted.

Really good audio was a long time in coming. In 1877, Thomas Edison had used a stylus that pressed a fine track into tinfoil wrapped around a spiral-grooved brass cylinder to make his first faint recording of a human voice. In the later system, to play back composition records on home phonographs, a steel needle fixed in a pickup head relied on the sounds produced by a simple diaphragm rattling to vibrations as the needle point bounced in the record's grooves. The resulting sound ran through a horn-like passage inside the phonograph cabinet and out a grille. The rattling diaphragm system was much the same principle that Alexander Graham Bell used in his telephone of the 1870s. That was to change.

The early microphones were actually modified telephones. After the electric microphone went into general use in about 1927, recording studios abandoned the big horn of the early recording system. In the new mike, electric impulses were amplified and fed into the recording stylus. Broadcasters instantly adopted the electric microphone. These early mikes can be seen in 1930s movies, to the amusement of modern generations. The real things are in museums.

The microphone pickup was suspended in the centre of a metal ring on fine springs to soften vibrations. Still primitive, it was an important step along the road to high-fidelity stereo audio. The improvement in sound was plainly significant. Previous recordings, long tinny-sounding to the comic extreme, became instantly quaint and obsolete. Only when the later hi-fi age came in the 1960s were they re-processed into revered examples of Jazz Age and 1930s musical masterpieces.

At first, radio stations broadcast just recorded music and brief, clipped announcements of the next musical number. The inevitable commercial "spots" began to appear. Suddenly, the prospects of actually *making big money* through radio reared up. Station owners began to see that there needed to be more program substance to attract audiences, who would then also listen to the commercials and presumably rush right out to buy the products.

All over the continent, live programs of every kind began to radiate out of the small local stations, who were still groping for the right formula to ensnare bigger audiences. They tried using live performers, but often the necessary studios and musicians were too expensive. Programs with children's voices dependably attracted local audiences. And the children didn't need to be paid. Their kids just being on the radio was reward enough for the parents.

Your author was, coincidentally, a member of one of those local children's shows. So, if personal submersion in the early days of broadcasting is needed for bringing Canadian radio history into focus, the following surely is enough accreditation. Before I was unwillingly plunged into the minor league of radio broadcasting at the age of seven, much pioneering had been done by risk-taking business adventurers who'd helped push the new medium beyond the crystal set to the more sophisticated batteryless radio.

In my hometown of Brantford, in Southern Ontario, circa 1930, a Mr. Brown was operating a primitive broadcasting station called 10-BQ out of his house on Terrace Hill. It likely put out little more than twenty watts. As my father Harry was an avid amateur radio-builder, my brother Dave and I could hear 10-BQ on a crystal set he made for us. This was my first encounter with the magic of radio.

We listened transfixed by sounds on the set's mysterious conglomeration of earphones (which I still have) a copper wire coil tightly wound on a cylindrical cereal box, a battery, a small crystal, and a tiny swivel gadget for manoeuvring a very short, hair-like piece of copper wire, or "tickler," to scratch the crystal and pick up a static-filled radio signal. There was a thrill in hearing something ethereal in the faint sounds of music and human voices that the tickler evoked. And there was a delicious twinge of adventure in the chances of actually tickling the right spot on the crystal to bring in 10-BQ, or some other nearby primitive station.

I also still have a tiny, two-inch cathode ray tube and some assorted oversized vacuum tubes left over from my father's attempts to build television receivers. In 1935, he started to build a mechanical set along the lines of that used by his fellow Scotsman John Logie Baird of Glasgow, the inventor of the first successful commercial TV system. In about 1936, Harry's TV actually did produce a rudimentary picture from the experimental television station KDKA in Pittsburgh. His device involved some vacuum tubes, coils and whatnot, plus a large cardboard wheel with little square apertures cut around and just inside its edge.

A scanning disc with perforations spiralling toward its centre had been created in 1884 by German physicist Paul Nipkow, but the necessary assistance of radio was not then available, and Nipkow's disc didn't develop. Research by others before 1900 also laid the groundwork for TV. In 1908, Scottish inventor A. A. Swinton suggested cathode rays as a television tool before the CRT was invented.

Harry Murray emulated the Baird mechanical system. An electric light bulb hooked to his receiver flickered in response to the TV signal picked up on a crude antenna. This flickering light was projected through the apertures of the spinning wheel onto a sheet of white paper on a wall, reproducing what appeared to be a televised object. I believe this object was a doll seen by a KDKA camera. With a similar system, Logie Baird had launched the world's first commercial TV transmissions in London, even in colour, in 1928. Under BBC support, he actually initiated a service in which a few dozen Baird TV sets were bought by the affluent. Regular telecasts were well under way when World War II broke out, killing them off for the duration.

Growing up in Brantford was probably, for me, an unconscious introductory course in communications. This may have been responsible for my later excursions into almost all forms of communications, from printing to radio, military wireless, newspapers, magazines, recordings, photography, and books. Others of my hometown contemporaries also found careers in radio and newspapers, even TV. The Telephone City, as it was known, was famous as the one-time home of Alexander Graham Bell, and the great inventor's former presence was, in those days, impossible to ignore, which may not be the case today in this age of now-ism. Yet the monuments to Bell, testimonial and physical, remain.

Brantford was also the home of Thomas B. Costain, internationally famous author of landmark novels such as *The Black Robe*, *Son of a Hundred Kings*, *The Silver Chalice*, *The Tontine*, and many others. He began his writing career as a young reporter on the *Brantford Expositor* newspaper during the First World War.

Another internationally known author was Sarah Lee Duncan, born early in the twentieth century, who grew up in a prosperous Brantford mercantile family. Her novels made her a darling of the international literati. She lived much of her life with her husband in India. At least one of her books was still being sold in 1984.

Lawren Harris, a Brantfordian born in 1885, became one of the famous Group of Seven Canadian artists in the 1920s. Yet another Brantford native was James T. Hillier, co-inventor of the revolutionary electron microscope. And, as the world knows, Brantford is the home-town of the king of hockey, Wayne Gretzky.

As school children, we heard all about Alexander Graham Bell and how he'd invented his revolutionary device while living in Brantford. Bell, of course, had nothing to do with the invention of radio, but his telephone was the earliest electric voice communicator to be heard by the public. On Sunday school and other picnics, we tots were often transported to the grounds of the Bell Homestead on Tutela Heights Road, still kept today as a shrine to Bell's achievements.

"Under yon roof the telephone was born," Bell himself said of his old home in his Scottish accent when he was the guest of honour at the 1922 dedication of the impressive Bell Memorial in the heart of Brantford. As he said it, he gestured in the direction of the far-off homestead. Though the physical development of the telephone took place in Boston, where the money was, its concept, its dream, and the young Bell's earliest experiments happened at the Bell Homestead.

We young picnickers, and later bicycle riding scavengers of the 1930s, often spent afternoons scouring the grounds around the Homestead for any bits of wire or pegs that might have survived the 1870s quest by Bell for a workable voice communicator. We'd long before been beaten to it. Elsewhere in Brantford itself, the Bell Memorial, a great wide stone and bronze panel sitting in its own small park, depicts a mythical passing of the spark of communication from the hands of the gods to the hands of humans. The oversized bronze figures were sculpted by Walter Allward, the famous Canadian sculptor who created the monument to Canadian soldiers at Vimy Ridge in France, where the Canadians won one of the First World War's most crucial battles.

Not only did Bell formulate and deem practical with primitive experiments his telephone idea at Brantford in the 1870s, but he frequently returned to the city from Boston for rest and relaxation. Then in 1879, two years after making his successful demonstration of the telephone to a group of flabbergasted Boston businessmen, he made the world's first long distance telephone call from a shop in Paris, Ontario, to an office in Brantford, seven miles away. That helped to seal forever Brantford's fame as The Telephone City, the place where the telephone was conceived.

war of the airwaves

Already, by 1932, there was a war brewing between the entrepreneurs of commercial private radio and the philosophers who wanted the entire world of radio broadcasting operated by government only, as ably recounted in Knowlton Nash's book *The Swashbucklers*. The now-famous Canada-wide radio and TV network the CBC hadn't yet appeared. Something called the Canadian Radio Commission held sway over all radio. I remember its logo still, on the studio door of CKPC Brantford, where I cut my radio teeth.

The fight for, on the one hand, freedom to broadcast any way and any how the private stations desired, and on the other, government rules on what might be okay to send out, was just starting to heat up. As a minutely junior player in those early days, I not only had no idea of a battle going on, but I didn't give a damn, either.

My *first* radio career, lasting from the age of seven to the age of twelve, emerged under the wing of a remarkable lady, Miss Josephine Whitney. Miss Whitney strove futilely for some time to teach me to play the piano. Then she decided I would be a good singer and interlocutor to add to a group of children she was organizing to put on church concerts, singing and giving recitations.

As far as I could tell, I had the singing voice of a frog. Whether Jo Whitney was really avidly religious or not I never knew. But she could play the piano and belt out Sunday school songs in a sort of controlled shriek. When we appeared live on stage in church halls we were a smash.

Miss Whitney employed this talent to train and encourage her group of a dozen boys and girls under the age of ten in a raggedy chorus of childish voices. Nobody but parents and those who doted on the very young would have excused the jagged sounds the group emitted, but it was enough to get Miss Whitney's proteges on stage. The intrepid Miss Whitney then convinced CKPC to put what she called "The Sunshine Kiddies" on the air. Consequently, in 1933, CKPC opened a slot for us on its Sunday morning schedule for fall, winter, and spring.

This radio miracle happened in Brantford, twenty-eight miles west of one of Ontario's main centres, Hamilton, and about seventy miles west of Toronto, the *big* city. Brantford was long established as a small, self-contained, and prosperous community with a dual function as both a farming and an industrial centre. Industry and countryside came together mainly through two major Canadian farm implement manufacturing companies, Massey-Harris, Inc. and the Cockshutt Plow Company, as well as the Waterous Boiler Works and the Johnson's Wax Company, with large plants employing over half the population. All these plants are long since gone, their huge factory buildings completely vanished. All, that is, except for Johnson's Wax which, coincidentally, was the sponsor of the popular weekly U.S. radio show "Fibber McGee and Molly." Brantford received credit on U.S. radio at the end of each show.

Offshoots, in the form of smaller steel fabricators, dotted the city, distantly related to the major steel companies in Hamilton at the head of Lake Ontario. Our church-hall audiences were mostly the workers and their families, with scatterings of farm families thrown in. Brantford was for a dozen decades a thriving hub of no small commercial import. Then, in 1930, The Depression fell on it like a hundred tons of steel.

However, life on the dole went on. As a first stab at cracking radio, Miss Whitney managed, in 1932, to place two or three of her young performers, including me, on "The Denver Sandwich Hour," a program broadcast daily over CKPC after school was out. This was the day of child-stars Shirley Temple, "Our Gang," and other beloved performing movie toddlers. Childish voices were in vogue on radio and on the silver screen.

Many later movie stars were yet to graduate from radio. Bing Crosby and Jack Benny were just getting airborne. Radio performers were held as objects of some awe, and were a sight to behold in the flesh. Jo Whitney's little crew had our own local following, but forever waited in the wings, never making it into big-time radio or the movies. After all, Canada would not become Hollywood North for many decades.

"The Denver Sandwich Hour" was sponsored on CKPC by the Patterson Candy Company, who made the Denver Sandwich candy bar right there in Brantford. The program, before a live audience of children crammed into CKPC's single studio and sitting all around on the floor, gave kids a chance to sing or do recitations on the air if they wanted to. Everybody who attended got a free Denver Sandwich bar. In 1932, that guaranteed a packed live audience every afternoon after school.

We Whitney performers had no say in whether we chose to perform or not. Miss Whitney simply told us to, and, with the reluctant obedience of children, we did. Shaking like a leaf, I awaited my doom in a corner of the studio while other kids did their stuff on-mike for the Denver Sandwich. In the control room behind a double layer of plate glass, a harassed Hugh Bremner, sole CKPC announcer, announced the next performer.

When he hit the words, ". . . and now, *Gilbert Murray* will sing 'Little Man, You've Had A Busy Day'!" I froze. My heart plunged, then leaped, then plunged again. Miss Whitney, sounding chords on the studio piano, gave me an insistent nod and I stepped up to the microphone. It was the typical 1930s-style mike with an electric pickup in the centre of a steel ring, hanging from an overhead boom so little feet wouldn't trip over a base.

I got out my first quavering words of the song, widely popular at the time. Thankfully, there was no such thing as tape recording then, as I would have been mortified forever if a recording had survived that session. Phonograph recordings were made only in the Big Time. Somehow, doing my best to keep up with Miss Whitney's piano, I got through the piece. For this bravery I was to endure unending humiliation, physical and mental, in the Victoria Public School yard ever after at the hands of merciless fellow pupils.

The Sunshine Kiddies' name didn't earn me any high level of appreciation in the schoolyard, either. I think teachers and my peers held the view that I was getting too big for my britches and needed downsizing. Fascination with upstart radio celebrities was totally foreign to the Victoria School yard.

But I did get my free Denver Sandwich bar.

A skillful promoter, Jo Whitney persuaded CKPC to put the entire complement of her precocious young choristers on the air. Our repertoire was mostly children's Sunday school hymns, mixed with novelty numbers — some written by Miss Whitney. An announcer was needed to introduce the numbers. For some obscure reason, Miss Whitney designated me for the job. By then I was eight. I held the exalted position of Sunshine Kiddies announcer.

Thankfully, the program was only fifteen minutes long. On the signal from Hugh Bremner (later to be locally famous in London, Ontario, as a favourite CFPL television newscaster) behind the double glass in the control room, the group struck up the rousing theme from Miss Whitney's own hand, "If you think that you want to be haaa-py /When you think that you're ree-ally sad . . ."

The newly-cast radio announcer, Gilbert Murray, read out the introductory script written by Miss Whitney welcoming the devoted audience to "The Sunshine Kiddies" program. Then followed solos, choruses, and recitations by the assorted child stars, introduced by the intrepid, but petrified with fright, young announcer.

When it was over, the sweat dried on my brow, Mr. Bremner signed us off, then congratulated us through the intercom.

The phenomenon of "mike fright" doesn't seem to have any basis in reality. The fact that a great number of people far beyond the microphone might be really listening doesn't justify the degree of emotional paralysis that it generates among some performers. Why should speaking in front of a microphone make any more difference to your inner composure than speaking to only three or four people in a room?

Perhaps it's the gut-feeling that if you make a mistake you are going to be pilloried and humiliated by far more people than by that roomful of fewer than half a dozen. But then, judging from the universally cool, composed manner of today's radio and TV personalities, it may have been something that only those who grew up in the confusing and perilous times of the 1930s suffered from. Or perhaps it was because tranquilizers hadn't been invented then.

The fact is that most people out there in Radioland are not hanging on to your every word as much as you think they are. Still, many performers on radio, stage, and screen have confessed that whenever they were about to perform they were so tense that they thought they'd never make it. It doesn't matter whether it's radio, TV, or movies that's involved, the sensation is the same. Usually, there is no basis for fear. The performance will probably come off without a flaw.

But there remains this icy feeling gripping the abdomen that you have committed a catastrophic goof, or are about to — a shame that will follow you forever through life, wherever you go.

I never got over that effect while announcing "The Sunshine Kiddies." Being on radio was bad enough for my schoolyard popularity, but performing on stage before a mass audience who were waiting to be entertained was devastating. You don't want to dampen your audience's great expectations. Despite my paralytic feelings, The Sunshine Kiddies were hailed as a great group of hometown child performers. Their puzzled announcer got due acclaim. I could never understand why.

Decades later, at a cocktail party in London, Ontario, I overheard Hugh Bremner, by then the local TV news superstar, recall how this youngster delivered his lines on CKPC with a professionalism like no other child announcer he'd ever heard before or since. He didn't know I'd overheard him. I slipped away. The mental paralysis of those 1930s days swept over me again.

depressionus exultatus

Without the movies and radio, it is hard to imagine how most of the North American population would have pulled through the Great Depression of the 1930s. The U.S. radio stars — Jack Benny, Bing Crosby, Bob Hope, Eddie Cantor, Edgar and Charlie, Fred Allen, and Fibber McGee and Molly, as well as the CBC's "Happy Gang" — in a desperate decade propped up a light-hearted view of life that, for many, had little hope and little real expectation of a return to prosperous times.

The actualities of real, everyday life were too hard to face without some fantasy to fall back on, in contrast with the forced hardships sought out by the advantaged youthful populations of the late twenti-eth and early twenty first centuries as a relief from prosperity. The choice of retreating to some exotic world where hardship seemed to be a refreshing novelty was not available in the world of the Depression. On radio shows and in movies, comedians hailed the noble example of penniless common folk bearing up despite their forlorn world. If they hadn't, laughter would surely have vanished from the face of the earth.

Except for Bert Pearl's "Happy Gang," there was nobody yet audible in Canadian radio capable of equalling their positive effect despite efforts by government-sponsored bodies like the CBC to encourage home-grown talent. Canadians stuck to their radios to hear the hilarious doings of the U.S. comedians. In self-defence many private Canadian stations became affiliated with U.S. radio networks, who carried the funny people onto Canada's airwaves.

A major privately-owned Canadian station, CFRB Toronto, valiantly tried to encourage a homegrown stable of talent in the 1930s and 40s, even into the 50s. The great comedy team of Johnny Wayne and Frank Schuster, eventually adopted by Ed Sullivan on TV, got a strong hoist into their ultimate Canadian and U.S. careers in the old live broadcast theatre of radio station CFRB at 37 Bloor St. West, Toronto, a once-well-known address, now long obliterated by a mas-sive office- and commercial-building complex.

Many other Canadian entertainers who ultimately made it big in U.S. radio and movies got started on "RB," among them singers Shirley Harmer and George Murray, Gisele MacKenzie, and Denny Vaughan, to name a few famous stars of the time. Americans Cary Grant, Buddy Rogers, and Red Skelton also were among the perform-ers who passed through the old CFRB studios. Another Murray, Jack (unrelated), produced a popular money show, "Treasure Trail," before a live audience in the theatre studio. Former *Toronto Star* reporter Claire Wallace, later famous on the CBC, started in radio with her commentary show on CFRB.

Veterans of earlier times in radio across the land could probably relate zany stories of adventures, both humourous and desperate, when things

Demonstration of the first successful voice communication system, the telephone, by Alexander Graham Bell in 1876 threw open the way for people to talk to each other even if they were on opposite sides of the world. A quarter of a century later, Reginald Fessenden built the first radio transmitter—also capable of sending voices, but without wires. The famous Bell Homestead in Brantford, Ontario, was where Bell did his basic theorizing and earliest experiments. Though he moved to Boston for financing of his project, he returned often to his parents' homestead for many years after. In the heart of Brantford is the magnificent Bell Memorial, its figures sculpted by Robert Allward, the internationally famous Canadian sculptor who created the monument at Vimy Ridge, France, in memory of the fallen Canadians of World War I. The Bell Memorial was dedicated in 1922, with Alexander Graham Bell himself present a few months before he died. In addressing the audience, he waved his hand in the direction of the Bell Homestead and said: "Under yon roof the telephone was born," confirming that Brantford was the location of the telephone's conception and earliest development.

Bell Memorial

Alexander Graham Bell

We owe our ability to hear the sounds of music and voices from distant points as though they were in the same room to a remarkable Canadian, Reginald Aubrey Fessenden. His was the first voice to be heard by anybody else at a great distance without the help of wires. Reginald Fessenden invented vocal radio. Just as Alexander Graham Bell was the Father of the Telephone, Fessenden was the Father of Radio, sending his voice by wireless radio 1.6 kilometres from Cobb Island in the Potomac River, not far from Washington, D.C., on December 23, 1900. An assistant, Mr. Thiessen, the first person to hear a voice sent by radio, instantly confirmed by telegraph that he'd heard words come over the primitive AM radio receiver. Six years went by before Fessenden broadcast the first radio program, Christmas Eve, 1906.

"Ted" Rogers, Sr., in 1926, created a new means of radio receiving by inventing the power vacuum tube. The new component did away with the cumbersome leaky storage batteries previously necessary for home radio sets. North American radio manufacturers were slow to adopt the new Canadian vacuum tube because of the need to re-design their radios, but by 1930 the changeover did take place, making the home radio set a more popular household feature.

An example of early Canadian radio stations was CKCK, Regina, launched on-air in 1922 by the Regina Leader, later the Leader Post. In 1928 the family of Sir Clifford Sifton bought the Leader, its home seen here in 1922 with transmitting towers on the roof.

CKCK's control room was typical of the pioneer radio station. Note the gramophone on the left, with hooded microphone placed at the speaker grille. This was a crude means of getting music on the air, and it picked up unwanted sounds, such as street noise. Two extra gramophones stand ready on the right. Also notable is the row of large vacuum tubes on the control panel, and the array of storage batteries below.

Pianist Al Smith performs in the CKCK music studio in 1935, typical of most radio stations of the time. By then, studios were stylish and soundproof. Acoustics were better-served by sound insulation, and studios were more photogenic. A floor microphone on the right serves the announcer. Unfortunately only his script is visible in the picture.

CKCK's control room by 1935 was equipped with a record turntable feeding its music directly into the system. The control panel had shrunk to desk-size.

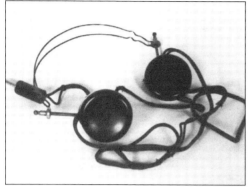

This set of earphones from around 1925 recalls the day of the crystal set when they were needed for hearing the weak signals picked up by the primitive radio. Magnetic loudspeakers appeared around 1930, together with more powerful detectors and amplifiers, enabling whole groups to enjoy together the hilarious antics of Jack Benny and other comedians and, for the first time, a better-quality sound of music.

TV sets with picture screens as small as this 1935 cathode ray tube's wouldn't have been popular when television sets emerged in the late 1940s. They would have been out of the picture, so to speak, as TV screens grew bigger in the 50s and 60s. This eight-inch long tube with a two-inch wide screen was for experimentation. The vacuum tubes dating from the 1940s were smaller than their predecessors', and the even smaller "peanut tubes" helped make possible small portable radios.

Disc players for radio broadcasting and home record players used fifteen millimetre needles in their playback heads. After each play of a record the needle was changed to counter wear and tear on the grooves of the ten- and twelve-inch 78 rpm discs and the rarer fifteen-inch predecessor of the LP. Pictured are needles and their tin boxes, two of them dating back to the 1920s.

Free-standing radio cabinets were part of living room furnishings by the mid-to-late 1930s and well into the 1940s. The floor models were crafted as fine furniture and displayed as a sign of growing family affluence. This 1938 Philco would have cost $100 or less, in 1938. AM only, it included a shortwave band able to pick up shortwave signals from around the world. Some of the rantings of Adolf Hitler were heard on this radio, as were the declaration of war on Germany by Britain's Neville Chamberlain in 1939, and that on Japan by U.S. President Roosevelt in 1941. It brought daily wartime newscasts, including those delivered by the booming voice of the CBC's Lorne Greene.

Two old-time announcers of the wartime mid-1940s take a break and get out in the sun away from their windowless studios. Left is Norm Bailey, with Gil Murray on the right. They were announcers on CKTB, St. Catharines, Ontario, during 1943–44, seen here just before Murray enlisted in the Royal Canadian Signal Corps.

This vacuum tube radio was part of a 1951 RCA "home theatre" that included a sixteen-inch TV, a 78, a 33 1/3, and a 45 rpm record player. It was among the last of the AM-only receivers to be produced for regular trade. A double-doored console cabinet with one speaker, also among the last of its kind, contained the components.

Floor model radios were going out of style in the 1950s, possibly because World War II veterans and their new families were living in smaller postwar accommodations. Table radios like this German-made 1955 Graetz Komtess were popular. It had AM/FM and two shortwave bands, even though there were few, if any, local FM stations.

During the nuclear war scare of the early 1960s, people were advised to have a battery-operated radio like this 1961 Sony on hand in case electrical systems were destroyed and survival instructions were broadcast. Ironically, this brought back battery-operated radio systems made obsolete in the 1920s by Ted Rogers' power vacuum tube, but using the small flashlight batteries developed since then. This time the battery-run radio stayed, and portables shrank to wallet-size.

The Liberal Party of Canada's 1968 Ottawa convention that chose Pierre Eliot Trudeau as party leader and prime minister was heavily covered by radio, TV, and newspapers. An example of a news team covering the heavily contested bid to succeed Lester B. Pearson was that advertised by CFRB, which like other stations across the country sent a full complement of news people and technicians to do remote coverage from the Ottawa convention floor.

As 1965 President of the Ontario Press Gallery, Gil Murray addresses the entire membership of the Legislature at the annual Speaker's Dinner, the first radio-man to do so. Left is Lieutenant-Governor Earl Rowe, Mr. Speaker Donald Morrow, and Premier John Robarts.

When CFRB moved its operations from Toronto's Bloor Street in 1965 to brand new quarters at St. Clair and Yonge it was, for a time, the last word in radio broadcasting facilities. Left is one of the control rooms in the old location depending on LPs and tapes.

This is the FM control room at the new CFRB quarters, long before Compact Discs appeared.

Competition between media for news listenership was growing in the mid-1960s, and radio stations were increasing their advertising and promotion activities. Above is a CFRB roundup of its by then extensive news staff.

Radio's Reach To The Stars

CFRB's "Early-Morning Man," Wally Crouter, set a radio longevity record of fifty years on the air. Until his retirement in 1996, he was by far the most popular daily radio personality in Canada, among listeners from all walks of life. A top-level golfer, he played with Bing Crosby, Bob Hope and Perry Como. Over his half-century radio career he interviewed and associated with the famous in entertainment, politics, and royalty. He was well known for his relaxed charm, wide knowledge, and sharp humour.

Famous Canada-wide for her part in the longest-running Canadian TV program, "Front Page Challenge," Betty Kennedy also was a star personality as CFRB's popular afternoon interviewer. In twenty-seven years she is said to have interviewed twenty-five thousand guests, most of them internationally famous, as well as every prime minister of Canada since Louis St. Laurent. Extensively active on many corporate and educational boards, she was appointed an Officer of the Order of Canada and in 2000 became a member of Canada's Senate.

Once known as "the biggest name in Canadian Radio," Gordon Sinclair became famous for his iconoclastic approach on his CFRB "News and Commentary" and "Let's Be Personal" programs. His views often earned him enemies, but he considered that part of his general appeal. Early in World War II his comments on Canada's military caused a top-level furor resulting in his being banned as an official war correspondent. A *Toronto Star* reporter at age twenty-two, he travelled the world in the 1930s as a roving correspondent. He was also the author of four books. As the most controversial broadcaster on Canadian radio, "Sinc" drew huge audiences on CFRB. His longtime exposure on the CBC-TV's "Front Page Challenge" made him a national figure, and his views on U.S. politics made him known in the U.S. as well.

Another voice familiar in radio, not only to his vast CFRB audience, but also to hockey fans as the between-periods announcer of the Esso commercials, Jack Dennett's was otherwise the reassuring, no-nonsense Sound of The News. His eight a.m. and six-thirty p.m. Monday to Friday CFRB newscasts seemed to counter most of the news crises in his calm, factual, and firm baritone, belying the actuality of his uptight nature. He left no doubt of his conservative views of daily happenings and drew audiences of three to four hundred thousand with each of his daily newscasts.

In 1928, CFRB's first announcer, Wes McKnight, in voice and name became synonymous with sports reporting in Canada and from the U.S. He was doing on-air news reporting when such broadcasting was rare, describing live the historic visit of the British airship R-100 over Toronto in 1930. He is believed to have done the first live report of a golf tournament. He interviewed a French wrestler who didn't speak English, so Wes asked and answered his own questions. A sportsman himself, he played tennis, golf, and baseball. A leading sportscaster in Canadian radio, for thirty years he reported the play-by-play of the Toronto Argonauts and Grey Cup games. Wes became CFRB's program director, and later general manager, from 1959 to 1966.

Every evening, Monday to Friday, during the 1930s and 40s, Jim Hunter dominated the Southern Ontario airwaves with his six-thirty p.m. newscasts on CFRB. Thousands turned up the radio at the call of a posthorn and the beginning of "A-Hunting We Will Go!" He first went on the air on CKNC, the old *Toronto Mail and Empire* station, moving to the *Toronto Telegram* later as radio columnist. He gained fame nationwide when he covered the 1936 mine cave-in at Moose River, Nova Scotia, for the *Telegram* and CFRB, in competition with J. Frank Willis of the CBC, and was credited with a world scoop when two men were rescued.

A *Toronto Star* staff reporter who was routinely assigned to do sportscasting rose to fame with one of the most familiar and compelling voices in Canadian radio. Foster Hewitt, just following his editor's orders, did sports reporting on the *Star*'s radio station in the late 1920s. He was the pioneer of what became on-the-scene radio sportscasting with his weekly play-by-play NHL hockey broadcasts and his trademark "He shoots! He scores!" When the Star's CFCA shut down in 1931, an advertising group continued Hewitt's show on CFRB. Later it was the "Imperial Esso Hockey Broadcast" on the CBC. In 1951, Hewitt founded his own radio station, CKFH, and, not happy with television, gave his CBC TV hot seat to his son, Bill.

Beginning in 1942, for ten years on CBC Radio, Claire Wallace was a nationally-known voice as the host of "They Tell Me," one of the first regular female stars on a radio network. Originally a *Toronto Star* reporter, she began in radio on CFRB in 1935 with an evening program, "Teatime Topics." On CBC, she had as guests a continuing roster of celebrities, including Dwight Eisenhower, and, it is said, Gene Autry's horse. Back at CFRB in 1952, she also was an author, advising on etiquette and travel, operating the Claire Wallace Travel Bureau, and fund-raising for charity.

The dominant voice on CBC radio during and after World War II, Lorne Greene became internationally famous as Ben Cartwright on the U.S. TV series "Bonanza," "Battlestar Galactica," and "Lorne Green's New Wilderness." During the war he was tagged The Voice of Doom because of his booming baritone delivery of war news both good and bad. He turned to acting after the war, running the Lorne Greene Academy of Radio Arts and performing on the New York stage. An appearance on TV's "Wagon Train" brought him to producers' attention and won him the fourteen-year "Bonanza" role that made him a TV fixture.

The Canadian Talent Library (CTL) came into being in 1965, putting recordings by Canadian performers on public sale. Funded as a non-profit service by Standard Broadcasting and forty other radio stations, CTL provided opportunities for Canadian musicians, composers, and arrangers to be heard on radio. Seen here in 1966 are George Harrison, President of RCA Records, and J. Lyman Potts of CTL, signing the distribution agreement for CTL albums.

didn't go the way they should have, especially in the days of all-live radio. Bill Baker, CFRB's longtime and original station engineer, told of a session he directed for the world-famous orchestra of Guy Lombardo and His Royal Canadians, inadvertently broadcasting what was probably the first simulcast in the history of radio. CFRB became affiliated with the CBS radio network in the 1930s, carrying live CBS shows and itself originating some to feed by telephone line to CBS in New York.

Playing a "gig" in Toronto at the time, Lombardo and his boys arrived at RB to do a live line feed to New York for their regular dance-band series on CBS, just at a moment when Baker's one and only amplifier was in full use on CFRB's own air, broadcasting a lucrative locally-sponsored program of recorded music.

This meant that, without an amplifier and its controls, the Lombardo orchestra's feed to CBS New York couldn't be audio-balanced properly, something the CBS engineers wouldn't tolerate. But the live feed was already near deadline. A survivor of many split-second crisis decisions in the years since CFRB's founding in 1927, Baker ordered the feed to be piped through the single amplifier being used to broadcast the station's own sponsored on-air recorded music program. He had the CFRB announcer say — on only RB's own air — that the next four numbers would be Guy Lombardo *recordings.*

As Lombardo started his band on its first musical CBS feed number live from an adjoining studio, Baker also switched it on to the CFRB air. That way he could surreptitiously balance the Lombardo feed right on CFRB's air as part of the locally-sponsored program, while also sending it down the wire to New York. The idea worked perfectly. Nobody but Baker knew the difference.

If during the live feed the orchestra had stopped for some reason, perhaps for Lombardo himself to announce a number, the game would have been up. As it was, "the sweetest music this side of heaven" went out over CFRB live simultaneously with the live feed to New York. The local RB sponsor was never aware that, instead of recorded Lombardo music, he'd had a free live performance of four numbers by one of the most popular dance bands of the day, at no charge. Nor did CBS know that the music they were wafting out across the continent was also floating over CFRB's airwaves at the same time, disguised as records.

Baker was proud of some of the other crises that he and his CFRB technical staff were able to put right. One was the aftermath of a tremendous rain and wind storm that knocked down the CFRB antenna at the transmitter then located at Aurora, north of Toronto on a back byway still known as CFRB Road. Baker and his technicians rushed to the transmitter and quickly devised a temporary tower to hang the transmitter mast on, and had the station back on the air in short order. Later, the RB transmitter was moved to Clarkson, on the Lake Ontario shore west of Toronto, for a big boost in power and a better signal.

Once more the Baker troubleshooters leaped into the breach when another disaster hit CKNX in Wingham, northwest of London, Ontario, near Lake Huron. The CFRB crew rushed new equipment to CKNX, a station that they had no actual connection with. They pitched in and got the Wingham station back on the air in little time. The principle of keeping radio forever riding the airwaves was at work.

Some of North America's most famous swing bands were broadcast live from "remote" locations around Toronto by CFRB during the 1930s — Duke Ellington, Glenn Miller, Tommy Dorsey and many others — with the resourceful Bill Baker at the controls. Live CFRB broadcasts of golf, football, hockey, and baseball, direct from various Canadian and U.S. cities, were regular features and part of its identity, rivalling the CBC network and swelling RB's audience by thousands.

CFRB's broadcasting of live talent was part of Canadian private radio's efforts to balance the effect of the steamrolling government-owned Canadian Broadcasting Corporation, with its mandate to run a network and give live opportunities for unknown Canadian performers. The CBC was also the federal regulatory body for all Canadian broadcasting, with absolute power over licensing of radio stations, and even home receiver sets, with strict rules for the licences.

Toronto and Montreal were the dominant centres of Canadian commerce in the nineteenth and twentieth centuries, with their large populations offering the best turf for commercial radio's success. Nevertheless, in the 1920s entrepreneurs bravely launched broadcasting operations in the less-populated Western provinces and Eastern Canada, creating the basis of today's successful stations in Winnipeg, Saskatoon, Calgary, Edmonton, Vancouver, Halifax, Moncton, and other cities.

A look back at those early decades shows a bewildering assortment of radio call signs, now mostly defunct. In Toronto alone the changes came thick and fast. CKGW, with five thousand watts of power, was founded in the King Edward Hotel in 1926 by Gooderham & Worts, the Toronto distillers, and split airtime with 1928-born CFRB on the same frequency of 910 kHz.

After leasing CKGW in 1933, the Canadian Radio Broadcasting Commission (CRBC, the Corporation's predecessor) renamed it CRCT, and moved its studios to the Canadian National Carbon Company's building. CKNC, the battery maker's station, whose quarters the CBC took over for CRCT, was shut down, but revived in 1936 as CRCY, again at one hundred watts.

The radio shell game went on. CRBC paid fifty thousand dollars for CNRT/CNRX, the Canadian National Railways Toronto radio operation, in 1933. A few years later, it became CBY, still a CBC station, then eventually in 1943 CJBC, anchor station for the new CBC Dominion Network. The Trans-Canada Network was run out of CBL. Much the same call-letter switching was going on all across Canada.

When CJBC Toronto went to fifty thousand watts in 1948, the CBC pre-empted CFRB's clear channel dial location of 860 for it and arbitrarily moved the big private station to 1010, where it remains today. Sixteen years later, CJBC became all-French. The Dominion and the Trans-Canada CBC networks had already merged into one.

Name changing had been going on in Canadian centres ever since radio began. The evolution of CKEY, a longtime popular Toronto station, is more complex, but is an interesting example of changes in the nature and status of a Canadian radio station. Long before it became CKEY, familiar to thousands of listeners, the licence was held by The Dominion Battery Company. As CKCL, founded in 1925 on a frequency of 840 kHz, it was assigned the same frequency of 580 kHz at five hundred watts as the Canadian Carbon Company's CKNC, with studios in Toronto's Prince George Hotel. At first, it split air time with the *Toronto Star*'s CFCA and Northern Electric's CHIC, a year later splitting it further with these two stations plus CHNC, CJBC, CJCD, CJSC, and CKNC. Confusing? There was more. Sharing time with other stations, now long since gone, went on with frequent changes until the Dominion Battery Company sold CKCL to Jack Kent Cooke in 1944, when it became CKEY.

The next year, CKEY was temporarily licenced to broadcast on five thousand watts in the daytime and one thousand at night. That went to five thousand watts full time in 1946. After Cooke lost his bid for the first private Toronto TV licence in 1960 and put his station on the block before he moved to the U.S., CKEY was picked up by Shoreacres Broadcasting, made up of the *Globe & Mail* newspaper, Canadian Westinghouse, Westinghouse Electric, and some Toronto lawyers. Westinghouse managed it. Eventually, Maclean Hunter Publishing bought-out Shoreacres and got an increase to ten thousand watts. CKEY went through many more metamorphoses. By 1991, it died and was reborn as CKYC, a country rock station. In 1997, it turned into CHKT, emphasizing Chinese content.

early in, early out

The *Toronto Daily Star* got into the radio game early. In 1922, the newspaper got a licence to set up CFCA at the old, *old* Star Building at 18 King Street West (the old building, built in 1928, was at 80 King Street West for over 44 years, its site now straddled by Toronto's tallest office building, at this writing, with seventy-two storeys, First Canadian Place). The *Star*'s astute guiding genius, Joseph E. Atkinson, as early as 1921 foresaw radio as something to pay attention to as a possible rival to newspapers for public favour, and apparently decided to join it rather than fight it. Eleven years later he killed CFCA, certain that the future of radio was with government public broadcasting. His foresight was faulty.

While it lasted, though, the *Star's* station was a limited success. After a — strangely, for the *Star* — unpublicized live concert of music over a telephone line to an audience at two Toronto locations, a series of *Star* free concerts was broadcast throughout 1922 over Canadian Independent Telephone Company lines. Atkinson had an antenna put up on the Star Building roof, and CFCA began its own broadcasting activity in June. It seemed that the *Star* was being well-established in the radio business.

In 1924, CFCA found new quarters in a building at St. Clair Avenue and Yonge Street — ironically the location, although in a new Procter & Gamble building, of Toronto's subsequently most successful radio station, CFRB, in 1965 after thirty-eight years on Bloor Street. By the next March, the future's great voice of hockey broadcasting, Foster Hewitt, was on the air over CFCA from the Mutual Street Arena. Hewitt, a sports reporter with the *Star*, fell into broadcasting almost by accident. As a humble *Star* reporter, he was routinely slotted-in to do the first hockey broadcast in Canada in the way *Star* newsroom people were always expected to take up any assignment they were given: no questions asked.

Hewitt had been reading news over CFCA and was the logical choice to do a hockey play-by-play. The fact that this hadn't been done before on radio didn't deter him or his boss. He squeezed into a glass booth down near the Mutual Street Arena ice and talked-up the three-hour game into a regular telephone hooked up to CFCA's control room. The nasal Hewitt voice of hockey was to be a hallmark of Canadian radio over the next several decades, even after CFCA went silent.

With Atkinson's retreat from radio in 1931, Hewitt's momentum only continued. MacLaren Advertising arranged the "General Motors Hockey Broadcasts," and CFRB took them over, along with Hewitt, and put them on a national private network patched together for the purpose. By 1934 the sponsor was Imperial Oil, but when the CBC was formed two years later, MacLaren's had to use the new national public network. The odd-couple of the CBC's CBL Toronto and CFRB were then joined together in broadcasting Hewitt and the "Imperial Esso Hockey Broadcast." Eventually, it became CBC Radio's exclusively.

When television arrived in Canada in 1952, "Hockey Night in Canada" was a natural for TV, and Foster Hewitt made a start with the new picture medium. Television wasn't for him, however. He eventually gave over the TV hot seat to his son Bill and stuck with his own station, CKFH Toronto, which he'd founded in 1951.

For awhile, Hewitt broadcast mid-week out-of-town Maple Leaf games over CKFH's power of two hundred and fifty watts. Ultimately, Telemedia bought the station and terminated the CKFH call letters. Hewitt died in 1985, and one of Canadian radio's most famous voices ceased.

The CBC set up the powerful Toronto station CBL as network anchor, airing live programs such as "The Happy Gang," and popular swing bands, but with heavy emphasis on classical music. Later, Andrew Allan, a onetime CFRB announcer who'd established an impressive reputation as a live theatre producer, founded a live CBC drama show, "Stage '44", followed by "Stage '45," "'46," and so on. CBC-TV was its nemesis after arriving in 1952. While they lasted, the "Stage" programs helped to establish and maintain CBC Radio's high-level cultural mission.

CFRB's venture in building Canadian radio stars faded and vanished partly because of the impossibly heavy non-commercial competition from the CBC. The cost of running radio stations was, by the 1940s, now colossal. Paying for live shows with available ad revenue was getting beyond reach. Anyway, the CBC was doing the job with its government-funding source. So, "leave it to the CBC" was the opinion echoed in private radio across the country. The era of more economic recorded music shows plus frenetic disc jockeys and newscasts then came to private Canadian broadcasting.

In 1941, the CBC appropriated CFRB's treasured clear-channel frequency, 860 — in the middle of the radio dial, to create CJBC Toronto, as a new Dominion Network popular-music-and-jazz CBC station. It assigned RB an obscure, less-clear one, 1010, farther up the dial. It was the third mandatory switch in dial location for RB since 1927. Postwar, CBC converted CJBC into a French-speaking Toronto station on CFRB's former clear-channel frequency.

In the 1960s, Ottawa deemed that the hand of the CBC on Canadian radio broadcasting was too heavy. It created the Board of Broadcast Governors (BBG). That brought the CBC itself under the supervizing control of an external body.

Radio in the 1950s became a magnet for eccentric personalities. Almost every broadcasting station from British Columbia to The Maritimes had its resident oddball, trapping audiences with either flamboyant on-air style or becoming well-known locally for mischievous antics. The microphone seemed to draw them. What better way to air their views than over the vast airwaves? They could project their off-the-cuff opinions on a wide, faceless audience and be known — sometimes loved and sometimes loathed — by people they'd likely never meet.

Once radio programmers recognized that the maverick personality could pull in big audiences, he or she became king or queen of the local ether, reigning over the airwaves, but under the station's command. It wasn't always so. Talk on radio up to the 1950s was usually short and quick. Before that, with some exceptions, few on Canadian radio were "personalities" of the kind commonly blasting out their egos over the air later.

Most earlier programming consisted of spinning 78 rpm discs of the classics or show tunes, with a little bit of jazz thrown in. Exceptions were when the "Big Bands" played in ballrooms and nightclubs in the larger centres. The more enterprising stations broadcast directly from the scene of the event.

In the 1920s, North America could tune in and hear the music of dance bands like The Coon-Sanders Nighthawks, as well as Paul Whiteman's, Roger Wolfe-Kahn's, Guy Lombardo's, and Ben Bernie's Orchestras, then famous everywhere over the U.S. radio networks, whose stations were easily picked up in Canada.

Broadcasts widened to even more on-the-scene "remotes" in the swing era of the 1930s, bringing into homes the music of swing bands such as Benny Goodman's, Glenn Miller's, Duke Ellington's, and Tommy or Jimmy Dorsey's, to name a few. In that era, live radio was a formidable rival of the movies. As the CBC's Elwood Glover said, all across North America people tuned in to the Coon-Sanders band, "The Knights of the Bath," every Saturday night for their live broadcasts from on-the-spot locations.

On into the wartime 1940s, Canadian stations were also broadcasting music of Canadian swing bands live. Among them were Mart Kenney's Orchestra, "The Best in The West," Trump Davidson's and Bert Niosi's Orchestras from Toronto, and others from Montreal. Fantastically popular U.S. swing orchestras were aired live from one-night stands at dance spots like the Palais Royale in Toronto and the Brant Inn in Burlington, Ontario. Through the early part of the Second World War, the airwaves were filled with the sounds of live orchestras. Then the need for bodies by the military sucked up most of the live talent.

A strike called by musicians' union chief Caesar Petrillo against the recording companies in 1944 paralyzed the recording industry for years. During the strike, radio stations couldn't get new records and, with radio being forced to turn to old recordings, the popularity of new songs couldn't develop. As true as it is of today, new hits were created when new records were played on the air. Orchestras, dependent upon hit record sales for their existence shrank or folded, putting thousands of musicians out of work. Caesar Petrillo was denounced all around as a musical dictator. Some artists like Frank Sinatra, using a choir backup instead of an orchestra, produced a few unmemorable releases. The final result was the demise of the Big Bands and the end of a lucrative era for musicians.

Having gained first-hand (or first-mouth) experience in the radio world of the early 1930s, later in 1943–44, and still later in the 1960s, I must now re-enter the story. Typical of radio novices, I was a high school dropout of 1941. This, as it turned out years later, was not due to scholastic incompetence so much as it was the confidence I held in easily turning into a "radio announcer" again at sixteen, after my five-year junior apprenticeship on CKPC. In fact, it was a regular thing in the late 1930s and into the 40s for high school dropouts to try for radio careers — a route that many stars were still taking in the 1970s and on. Some of us, after war service changed our career paths, got back to school after the war and completed our formal educations, even up to university graduation.

But in 1941, having kissed high school goodbye and confident of moving on into the bigger time, I lost no time in hitting up CKPC for a job. Strangely, although I'd seen myself as one of that station's "stars," the management spurned its chance to pick me up now that I was fully in the market. I was baffled at their lack of seizing on a good thing while it was going. When that happened, I then let the CBC know I was available for an announcing appointment. You shouldn't hide your light under a bushel, they say.

I got a nice letter back from Lorne Greene, the CBC chief announcer and wartime "Voice of Doom," later to be famous as the boss of the Ponderosa Ranch on U.S. TV. He let me know there were "no openings at present," but enclosed a government-type job application form. I quickly spotted the obvious disqualifier: What university degrees did I hold, and When and Where did I get them? Not being even close to a *high school* diploma, I threw in the towel.

Apparently, for the CBC of that era, actual broadcasting experience rated far below academic level for a "position." University education in those days was mainly possible only for the well-off. The sixteen-year-old offspring of cash-strapped Scottish immigrants wasn't likely to be among that elite. But I still couldn't believe these people would miss a chance to hire such seasoned talent. Oh well, the war was on and the army was biding its time.

I was seventeen and apprenticing in the printing trade when I managed to get hired on as an announcer with CKTB, "The Silver Spire" station, in St. Catharines, Ontario. Unlike the CBC, formal education as a requirement in private radio apparently didn't matter, especially with stations losing their on-air staff to the armed forces.

In 1943, when this event took place, there were only a handful of radio stations in Southwestern Ontario, my general bailiwick. I didn't even think of going beyond my home province. In Southern Ontario, outside Toronto, there were only private radio stations CHML and CKOC in Hamilton, CKPC in Brantford, CKCO Kitchener, CFPL London, CKNX Wingham, CKCW Windsor, CKTB St. Catharines, CKWS Kingston, and CHEX Peterborough. Today they are as common as telephone poles.

birth of a station

Because of wartime shortages of equipment, no new stations could be founded, and none of the existing ones could expand. All AM (audio-modulated) stations did have a broadcasting range far greater than the later FM (frequency-modulated) stations, but were low-powered and low-fidelity by today's standards. An Ottawa order froze all transmitting power. The Hamilton stations were putting out a thousand watts each, but most others were a hundred, the power of a mid-sized light bulb. In Toronto, anchor CBC station CBL and CFRB, the biggest private station, each fired out five thousand.

The Southern Ontario population in general seemed to feel they were well served by this assortment. However, there were also numerous U.S. stations available, broadcasting into Southern Ontario directly from Buffalo and Detroit and others over the CBS and NBC networks. It wasn't until the war was over that Canadian stations could raise their power. Ten thousand and fifty thousand watts became commonplace. Even later, with the emergence of FM on a large scale, power escalated to two hundred thousand and more.

CKTB in St. Catharines, Ontario, was typical of the smaller-centre stations. Its hundred watts covered a radius of about a hundred miles in all directions, with a freak intermittent effect that bounced its signal as far east as Nova Scotia and even, phenomenally, south to New Zealand. Other stations undoubtedly achieved this too, as AM radio signals don't penetrate the ionosphere, the Heaviside layer in the high atmosphere. They rebound to earth, bounce up and down again a few more times, and are receivable great distances away. FM signals, on the other hand, go right through into outer space and don't bounce back.

CKTB would often get official postcards from "ham" (or h-amateur) short-wave radio aficionados at incredibly distant points (this was long before satellites), asking for confirmation that they'd picked us up, citing program excerpts and giving exact times of pickups. Though this might not be unusual for short-wave signals, it wasn't supposed to happen with regular long-wave AM. But happen it did.

Decades later, when I was with CFRB Toronto, on a trip to the east coast with the car radio fallen silent but still tuned to 1010 on the dial, Betty Kennedy's voice suddenly came in clear and strong as we breasted the hill over Halifax. Startled, I realized I was getting a "bounce" of my home station from the Heaviside layer. It was something to talk about when I got back.

My new 1943 employer, CKTB, had an odd beginning. The "TB" stood for Taylor & Bates, an old St. Catharines brewery that closed just before World War II. The "Silver Spire" was the brand of beer it made, and also described CKTB's tall, silvery transmission tower at Port Dalhousie, a few miles away on Lake Ontario. Edward T. Sandell, known by his initials "E. T.," owned both the Taylor & Bates brewery and CKTB.

The station began in 1930 as just a studio and control room in the Welland House hotel in the city's downtown. Programs were fed for thirty-five miles by phone line to CKOC Hamilton — a station that went on the air in 1922 and is still very much alive — and were broadcast from there. Many Canadian stations began as "feeders" into licensed broadcasting outlets until they got their own licences.

When CKTB was licensed in 1938, Mr. Sandell moved it to a huge nineteenth-century house perched on St. Catharines' Oak Hill, high above a disused 1880s Welland Canal branch. Away down below at water level lay the old ramshackle T&B brewery. An ancient tunnel, reportedly used temporarily as a prison for Mackenzie rebels in the 1837 Rebellion, burrowed up deep under the high bank and connected the brewery with the house. By the time I arrived there, a solid door in the big house's cellars was nailed permanently shut.

The historic Oak Hill house — already in 1943 well over a century old, and which may have been the mansion of William Hamilton Merritt, builder of the original Welland Canal — was renovated for radio broadcasting. Installing what was then the latest twentieth-century electronic equipment in an early nineteenth-century house was indeed an incongruous historical switch, no pun intended. Two studios, one large and one small, were built on the ground floor, the rest of the floor being offices. E. T. Sandell's luxury apartment took up the second, reached by the mansion's broad, sweeping, carpeted *Gone With The Wind*-type staircase.

The Oak Hill mansion was a gracious setting for a radio station. On the third floor, the servants' quarters in the old days, were two large rooms and a bathroom that were all shared by three announcers at a bargain rental. Huge circular windows more typical of the *eighteenth* century looked out in all directions, most notably over the old canal and civic gardens decorating its high, steep banks, giving our living quarters sweeping views. One could imagine the servant girls who must have done their wistful gazing out at much the same scenes a hundred years before us.

But there were no girls anymore on the third floor. This was where I was to live for the next many months. For the first time in my young life I could jump out of bed five minutes before working time, run down two staircases in my pyjamas, turn on the transmitter, read the day's first newscast on the air while catching my breath, and grab some clothes before the office girls arrived for work.

The station had an all-woman management. Marian Hallett was the manager and Genevieve Heibert was secretary/treasurer. Mr. Sandell seldom appeared on the ground floor. A tall, silvery-haired man of seventy-two, he would usually be seen only when passing through the lobby on his way to his car and the golf club. However, when I first joined the CKTB staff he personally took me on a tour of the mansion, proudly pointing out how it was built with "solid English brick" imported in the 1830s.

Today, radio stations project voices from air-conditioned high-tech studios that look more like parts of a NASA launch station. "Talk hosts" and disc jockeys and their guests, isolated in studios inside double plate glass soundproof picture windows, crouch over vinyl-topped desks at hanging mikes like spacecraft appendages, heads encased in clamshell headsets that feed their own dulcet tones back into their ears in what must be the most ego-satisfying process ever invented. Rows of coloured backlit buttons and sound-shaper slide handles lie arrayed in front of them, or control-room operators, in a jaw-dropping, arcane panorama like a Boeing 747's instrument board, lacking only a radarscope. Even a radar screen for instant weather reports may be there to fill that role. Banks of CD and cassette tape slots rise before them.

In the early 1930s and 40s, control rooms were also made acoustically remote from studios by heavy double plate glass windows and soundproofed walls, but the control boards in them looked primitive compared with today's. Strictly monophonic, the boards had a single row of black "pots," or control knobs (attenuators), over a row of toggle switches. Each controlled the sound levels of various microphones in the control room and the studios, and the pickup heads of the record turntables. We didn't wear headsets, unless to troubleshoot.

What went on at CKTB among its crew and general operations was probably typical of most Canadian radio stations of the day. Old hands would find all of the following examples of day-to-day adventures familiar. To make a profit, any radio station pursued a single goal: broadcasting entertaining and informative programming that would encourage listeners to go out and buy products advertised on-air, partly because one of their favourite stars said to. Not much has changed in that.

There were technical limitations on what a station could do. Because of the hot vacuum tubes, the equipment could easily burn out and had to be maintained, even by keeping stations off the air overnight while everything cooled down. In wartime, upgrading of equipment was near impossible and the status quo reigned. But daily operations were much the same everywhere. Pressures were the same. At almost all private commercial radio stations, the announcer of the moment was also the control room operator, charged with smoothly announcing musical numbers and reading short newscasts and commercial "spots," and also with spinning records on the two big turntables at our elbows. Years after my stint at CKTB, when I first joined CFRB, I was amazed to see announcers routinely working in studios while separate operators worked in the control room, spinning records and commercial tapes. What utter luxury!

In St. Catharines we made station breaks with the cry of "CKTB! The Silver Spire, St. Catharines, Ontario!" Listeners were expected, before the Taylor & Bates brewery shut down, upon hearing "Silver Spire" to put two and two together at a time when advertising beer was not allowed on Ontario radio stations.

The control room was, for me, a magical place. On the first day of my second radio career, with a little coaching and supervision, I was running the control board, cranking the pots up and down as needed, and riding the VU meter's leaping decibels. I was not a total novice. I did have a smattering of experience with things electronic. Besides my days observing CKPC's operations, I had a father who was a radio and TV tinkerer and, although he never told me much about how his gadgets actually worked, I was already demystified about all the knobs and switches. Now I was ready to roll.

bucking the bronco

But there still were some hard lessons to learn. With even just a hundred watts of power at hand, it was possible to advertise inadvertently a goof across a great distance and among a lot of people. I came closest to this through the antics of Norm Bailey, one of the two other CKTB announcers. Norm was a tirelessly gleeful prankster. He'd been through the wringer himself as a novice announcer, so he deemed me his deserving successor. All the older announcers who'd administered his own initiation were gone into the military, so he felt he had to carry on the tradition. A tall and cheery-faced jokester, something of a clothes-horse dresser, and a ladies' man, Norm seemed never to feel down.

An ambition of all announcers everywhere was to "break up" some hapless, still-green fellow on-air man. If you could make an announcer reading the news or a serious commercial spot collapse in helpless laughter while still on the air, but without the audience or management knowing what was going on, you'd have accomplished your main act of the day.

One of Norm's favourites was, while I was sombrely reading out war news, to rig up an impromptu clothesline in the little studio on the other side of the glass partition where I couldn't help but see it if I glanced up. On this he hung out a "washing" of toilet paper sheets with paper clips. Each sheet bore a crayoned letter of the alphabet, and in total spelled out some un-airable word. I would, theoretically, break up at the sight. And I usually did.

Then Norm would dance about in the little studio, grinning, wiggling fingers alongside his ears, and maybe sticking out his tongue. All of it was inane, but to the newscaster keyed up to addressing the control room mike with careful diction and sonorous tone, keeping in mind which commercial he was to read next and which turntable he was to spin after the newscast, this ridiculous charade usually proved to be too much. Anything . . . anything at all would do if the tormentor could just disorient his victim and cause him to guffaw, turn off his mike, and slam on a record until he could recover. It worked on me.

All this was at the risk of disrupting the station's broadcast "sked" and causing a breach in the written daily log of its time sequences. The result could be questions from the boss and a stern reprimand of the station from the all-governing CBC, to whom every minute-by-minute occurrence — good or bad — had to be reported, at the station's peril of losing its broadcasting licence. Horseplay wasn't on its list of permissible acts. As for the announcer victim, he was honour-bound not to implicate his prankster colleague.

Jack Dawson, longtime announcer and ultimately station manager at CFRB, told me about a dramatic incident that happened at that station long before I worked there. While Jack Dennett, a particularly serious major newscaster, was deep into his evening newscast, another member of the staff (Dawson is the chief suspect) sneaked into the small news studio, took up a position in a corner with his back to

Dennett, and from a great height very carefully and slowly dribbled water from a jug he'd brought with him into a drinking glass. Its source unseen, the gurgling sound was suspiciously like something of a very personal nature being done in a little private room.

Of course, the gurgling was quite audible on the air and also to Dennett. Embarrassed and furious, he turned off his mike, swung around, and was about to blast his tormenter when he spotted the jug. Still furious, but on the brink of laughter too, he yelled at the prankster to get out, then turned back hurriedly to switch on his mike, where, still quivering, he had to regain his normal calm and authority to go on with the news in his customary solemn manner.

The unfortunate control room operator, faced with accounting in the daily log for two minutes of dead air, threw up his hands. To Jack's credit, in spite of the dead-air gap, he came back on with his usual calm. The audience — perhaps intrigued by the strange watery background sounds — became none the wiser. The prank vanished into the annals of radio lore.

Another trick, played by another announcer, was probably the most outrageous. In this one, a perfectly reputable newscaster was reading the news over the air when a colleague slipped into the studio quietly and unnoticed. Everyone knew that a certain top station executive, who will remain anonymous, had a beautiful wife. The mischief-maker then leaned in to the newscaster's ear and spoke very loudly: "Okay, when did you stop sleeping with ********'s wife?" then quickly ducked out of the studio.

The newscaster, taken completely aback, stopped reading and looked wildly around him for the culprit, then through the window at the control operator. *Had the mystery voice been heard over the air?* Everything looked normal. The flustered newscaster then switched on his microphone again and resumed reading the news, somewhat quavery as his panic abated and his anger rose.

Afterwards, he quivered at the expectation that he'd be instantly fired. But nothing was ever said to him about the ghostly incident. Although he could guess, he never found out who the perpetrator was. In actual fact, his prankster colleague had set it up with the control operator to, at his hand signal, shut off the studio mike as he was about to speak, then quickly flip it on again. The operator, part of the plot, looked busy with records for the subsequent music program. He remained dead-panned and didn't even glance into the announcing studio.

My most harrowing close call at CKTB came with the help of Norm Bailey — *again!* This time he was an innocent participant. It was in the matter of my own grand major ad-lib boffo on-air *faux pas*. Here, a little explanation of the workings of the control board is in order: To go live on the air from the control room mike, its toggle switch had to be pushed into the "up" position.

To talk on the "talkback" to someone in one of the flanking studios you pushed the switch down. He or she could reply by switching on the studio mike. This was a vital fact for constant alertness. There was often a lot of talkback between the control operator and, say, a newscaster or a group in a studio waiting to go on the air, unheard by the radio audience.

This particular mid-morning, Norm was sitting at the mike in the little news studio, awaiting his cue. I was "on the Board." Having about five minutes before air time, Norm, a minister's son, set about entertaining me with some yodelling and singing of the kind of ditty that wouldn't be allowed on the air for at least another forty years, using his two-way toggle cough-switch for his studio mike talkback — correctly. But our high spirits were heading for a downfall.

While a record was on-air, I would work my mike switch and inject a risqué remark to Norm, and he'd risqué me back. Unfortunately, as you might have guessed, I'd absent-mindedly pushed my switch up — the on-air position — which, through some cross-circuiting, accidentally put Norm's mike on the air, too. Everything I was quipping at Norm and he at me was going out over the air. For several minutes I didn't realize this.

The time for Norm's newscast came. He went to air. I waited for the phone in the control room to ring or the station boss to appear. Nothing happened. I lay low and said nothing to anybody about my goof. Of course, in 1943, audiences were not as quick on the attack by phone as they have become since. Radio and its disembodied voices were often unreal to listeners, who seemed to believe that radio people actually lived on Mount Olympus. Few would telephone Mount Olympus or a radio station except in the direst of happenings. How times have changed!

When Norm got off the air, I confided in him my sin. He was shocked, but said to keep quiet about it. I did. Later, when we went for coffee and doughnuts at Murray Morton's lunch counter around the corner, we were still saying nothing and hearing nothing. Then Murray, serving us our doughnuts, asked what the hell was all that about that morning with all those voices chiming in over the music. We muttered something about technical difficulties beyond our control.

But Murray, whose lunch counter radio was always tuned to CKTB, gave us an arch look. He knew a little about broadcasting, being the leader of a part-time local dance band that sometimes played over CKTB air. But that was the end of it. Not a word came from the top about that odd mid-morning broadcast. Was anybody really listening? We chose to think they were just being kind and discreet.

It was at Murray Morton's lunch counter that I, as a new boy, was a subject of discussion about my status as an announcer at the age of seventeen. Somebody said I was just a kid but had the voice of a forty year old. I took that as a compliment. Then somebody else said I had the *right name* to be in radio, with the implication that this had some connection with how I'd been hired. That went over my head. Much later I learned that a former head of the CBC was a man named Gladstone Murray. In 1943, I'd never heard of him.

early cool

As time passed, inevitably there would be more fluffs and goofs. After all, an announcer-op had a lot on his hands. He must, while slipping the right records onto his two turntables, open his mike and chat calmly and coolly on-air about the music coming up, read a commercial convincingly, watch the clock, carefully slide a turntable start handle to "on" as he spoke, remember to roll down his mike's pot to avoid a clunk when he switched it off, and roll up the turntable pot to put the music on-air. Manual and mental dexterity were the rule.

Also, while a record was playing on one turntable, he had to change the steel needle (there were no LPs or microgroove cartridge heads yet)

Nothing On but the Radio

on the head of the other tone-arm after every play, slip on the next record, cue it with a quick slide in its leading groove ahead and back, and check the sked for what spot came next, all in the three minutes that a ten-inch 78 record played, then do it again for hours more.

All of the above hazards left massive room for error. Broadcasters took pride in running a radio station as though it were a goof-proof machine, avoiding dead air, unintentional talk, skipping records, and all the unwanted sounds that the air is heir to. Goofs were bound to happen, but were rare. That is, except for the time I was running the *Green Hornet* program one evening early in my new career.

The *Green Hornet* was a famous mystery/detective story in pulp magazines and on radio, featuring a fellow who drove a sleek futuristic super-car called the Green Hornet. He came to be identified also as the Green Hornet himself. Constantly in pursuit of criminals in his half-hour weekly radio program, he would put away wrongdoers in superhuman style, setting everything right. He was contemporary with Superman, his closest competition. Dedicated fans followed his exploits. Today's Spider-Man, Batman, and others are fictionally descended from these types.

Every week, CKTB received through express a large, flat package containing the current *Green Hornet* recordings, "bicycled" from some other radio station to us after they'd aired it. Other programs and commercial messages were also bicycled around like that. It was a miracle that the pair of large discs survived each delivery, as they were made of glass. The heavy, black glass discs, twenty inches across, substituted for earlier big discs of light metal, gone with the war.

These were early "long-plays," running at 33 1/3 revolutions per minute, just like the LPs of years later, carrying fourteen minutes of sound per side, allowing one minute in between for a spot commercial. The difference from the LP was that they were not microgroove records, but were played with the same steel needles as used with 78s running in the same kind of groove. Another difference was that they had none of the surface noise characteristic of the composition 78s.

The glass grooves were so smooth that there was absolutely no needle hiss to reveal over the air that the program was not live in our studio, but recorded. The heavy discs had to be handled with great care, lest the glass be broken and the next station down the line — as well as

56

us — be deprived of the gripping *Green Hornet* adventures. Experiments with 33 1/3 speed had been conducted as early as 1932, when recordings were made on them by Duke Ellington and Bing Crosby, among others. Those recordings have audio fidelity remarkable for the time.

Two *Green Hornet* programs were on the four sides. The first half of Program One was on Disc One, Side One; the second half was on Disc Two, Side One. That meant the operator laid Disc One on Turntable One, Side One up, and Disc Two, Side One up, on Turntable Two. The first fourteen minutes of the half-hour program would be played on Disc One, Side One. The announcer would air a one-minute spot and then, his hand on the start knob of Turntable Two as he talked, set off the second fourteen-minute half off Disc Two, Side One. I'd often run this show before with no hitch in the sequence. But not this time.

I was momentarily distracted either by a phone call or by some kind of puzzle in the sked, but when it came time to spin the second *Green Hornet* segment, an unexpected crisis developed. At first, busily shuffling sked sheets or 78s, I didn't notice it, not really listening to the program itself. Then the first of a string of phone calls came in from a listener asking what the hell was going on. Nothing was making sense in that night's *Green Hornet* mystery. What were those guys up to? Better go tell them they had things mixed up (the glass discs were doing their intended high fidelity "in-studio" stuff). What happened in the first half wasn't being followed up in the second.

The callers, more numerous as minutes passed, were right: I'd put the first segment on okay, but then somehow I'd put the first segment of the *next* week's program on Turntable Two, now playing. Listener confusion was guaranteed. In a way, the reaction wasn't a bad method of measuring audience listenership.

What to do? It was impossible to do anything. When the show ended, I read another spot and exhorted the audience to "listen-in next week for *more* exciting adventures of the *Green Hornet*." The phone call flood finally stopped. The next day, nobody from management asked for an explanation. Nobody seemed to know anything about it. Again, how different were the audiences of 1943! Today some irate listener would be on the phone to the boss, demanding the satisfaction of having the culprit fired, or threatening a lawsuit over the deliberate deception, or whatever.

My next major goof came a week or two later. I was running the usual recorded music and doing commercial spots as scheduled, taking over the Board from Ronnie Congden, the chief announcer, right after the six-thirty p.m. news. At seven, I had a program of classical music to air. Nearly all of the music played over CKTB — and most other stations — was on 78 rpm phonograph records.

I regarded myself as an old hand at these, having for years spun my own ten-inch 78s on my little record player at home. This program involved several twelve-inch 78 rpm Victor Red Label records. I think it was a Beethoven symphony that crossed me up. There was a total of six record sides on three discs to play. I started with Side One, giving an appropriate intro, then went on to all the assorted control room chores.

When Side One came to an end, I switched on the other turntable and got a smooth segue into the next record. But what came out was *Side Three!* I knew this instantly, because I'd suddenly noticed that on the backside of Side One was Side Two, which was supposed to be playing. I'd handled the discs in much the way I should have put the *Green Hornets* on the turntables: Disc One, Side One on Turntable One and Disc Two, Side One on Turntable Two. The difference was that the 78s' sequence went from Side One to Side Two on the flip-side of the same record — something I should have known, and did in fact know.

I called in Ronnie, who was a sometime musician and likely had a solution, and confessed (Ronnie, back in the 1920s, had been auditioned in Paris by Arthur Horowitz, but hadn't made the grade to concert pianist). Keeping his cool, he sorted out the records, recovered the ball for me, and that was the end of it.

A CKTB standard operation was the broadcast every Sunday morning from a local church. This was a job for Ronnie, being a pianist and organist and therefore, it was assumed, at home in a church. Ronnie had other attributes that didn't include being a churchman. But every Sunday morning he was on hand at the church with remote broadcasting equipment, ready to send the message of the minister out over CKTB's airwaves by way of a telephone line and the station's hundred watts.

Accomplishing this feat was the acme of flying by the seat of your pants. At an appointed moment shortly before eleven a.m., the control room operator, namely me, cranked a handle on the field telephone, which buzzed Ronnie's. Then it was a game of my reading the scheduled commercial and telling the audience we would now go to the church for an enlightening broadcast. I'd flip the right switch on the board and the music of an organ would theoretically start to drone over CKTB air. No, there *wasn't* a hitch in the procedure. It worked every time, and "The Word" was safely broadcast.

Sunday at CKTB was its date with survival: the Day of the Dollar. That was the day most of the week's revenue was earned. It came from two main sources: a string of paid ethnic music programs in the early afternoon followed by a string of paid religious programs on into the evening. There was a mixture of Central European programs, one immediately following another.

In each case, three or four people of such origin would arrive at the studio ten minutes before air time, bearing packs of twelve-inch 78 rpm records of their own cultural music for me to play over the air during their shows. They would do the announcing. With local advertising not quite bringing the station's revenue up to a profitable level, the ethnic and religious monies were vital for CKTB's survival.

This was probably not unusual among radio stations of those wartime days. With most manufacturing production directed at producing war *matériel*, the flow of consumer goods was at a low point, making radio advertising very expensive for many local stores and businesses, with resulting falling-off in shopping. The Sunday programs were a godsend. Without them, CKTB and many other stations would probably have gone down the drain.

For the announcer on Sunday duty — again me — the day would not be exactly a picnic. It being 1943, and the middle of the Second World War, there were frictions involving various camps. For the time being, the Soviet Union was one of our allies. Therefore, the Russian show was friendly.

Between other groups, however, there was unconcealed animosity because of split allegiances back in their war-torn home countries. When one particular group had finished their program and was filing through the main lobby, and a rival faction was just coming in and passing at close quarters, things, to say the least, were not serene. It was a lesson to me in ethnic frictions.

I'd always thought that, this being Canada, with everyone united against the Nazis, there would be a harmonious political blending of all groups, and that perhaps enthusiastic handshaking all around would happen as they passed in the lobby. Wrong. When that happened, the lobby air was as frigid as the sub-Arctic. In fact, on one or two occasions when shouting and danger of a fistfight broke out, I — a wisp of a seventeen year old — had to step out of the control room and bark a cease-and-desist order to avert what might have turned into an international incident over CKTB air.

Besides that, there was a federal government order that no foreign languages were to be uttered over Canadian radio stations on the chance that signals to Nazi sympathizers might be buried in them. So, when at times one of the groups' spokesmen got off a few phrases in his/her native tongue, I had to make threatening gestures through the control room window. When a record was on, I'd have to warn them sternly through the intercom that if they did that again I'd cut them off the air. That usually worked — for the moment.

The religious people were much more serene. The groups passed in the lobby with angelic expressions on their faces, regardless of how they might disagree on the Gospel. Trouble was more likely in the studio — on the air. These were all fundamentalists. One of them featured a young preacher so fervent that his voice rose to a multi-decibel shriek in his preaching, threatening to blow the station off the air.

Before his program began, I had to test him by having him move farther and farther back from the mike until his volume allowed the VU decibel meter on the Board to register an acceptable level. He would then deliver his tirade from about twenty feet back, half-crouched, Bible in his left hand and beating with his right arm to emphasize his message, even though the audience couldn't see him. He would have been great picture material ten to twenty years later on TV.

Another show, aired in the evening, was the domain of a clone of Billy Sunday, the famous U.S. network religious phenomenon who held huge live audiences enthralled in his evangelistic broadcasts. Our local Billy and his wife drove over every Sunday evening from Toronto in their Cadillac, with their three teenaged children in the back seat. All of the children played an instrument of one kind or another. On the air in our big studio, mother played the piano, son played the accordion, one of the girls the guitar, and the other also played the guitar and sang. Billy himself fervently preached his own kind of message.

All went well in this evening Gospel–musical until, on one occasion, Billy quietly inserted into his sermon the news that young Billy junior was in need of a new accordion. No more said. During the following week, money by cheque, money order, and cash flowed into the station, dedicated to buying Billy junior not only one new accordion, but two. Such an instrument ran into the hundreds of dollars.

Of course, the money flow would exceed the need. The surplus presumably went to fund other needs of the family mission. The duty announcer had to listen carefully during future programs for any solicitations more obvious than for new accordions. It was a sly game.

There was a pleasant side to the station's live programming. Mid-mornings, a singing pianist named Clarence Colton crooned to the housewives and, as they used to say, tickled the ivories. Sometimes I did the announce work in the studio, standing alongside Clarence's piano at a floor mike. This had to be a rambling, jovial kind of announcer chatter with Clarence's piano tinkling in the background. Clarence had almost no lung power. When he sang, or even spoke, the control op had to boost the VU meter far higher than normal to pick him up, then crank it back down when I came on.

Clarence was an amiable, laid-back, serenely-smiling performer with few words except those of a song, skilful in sliding into any requested popular number, old or new. The day of radio's live, rambling, singing pianist is decades-gone. But Clarence delivered a musical panacea that soothed mid-morning audiences as much as the knock-'em-dead rock bands did later. It was a pleasure to announce him.

In those days of live CKTB performers, Jean Beatty, an effervescent daytime young woman chatterbox, was another audience-pleaser. Jean, hardly twenty by then, predated most of the later female broadcast personalities with her bright style of comments and handy tips for women. Immediately after the war ended, Jean moved to New York City to do a local radio broadcast. There she fell into the company of some radical intellectuals. She then produced a best-selling novel, *Blaze of Noon*, which recounted through fiction the exploits of a young woman newly-landing amid the New York political intelligentsia. Her book did blaze for awhile, but she never duplicated the achievement.

There was hardly a halt in the cycle of personalities. A daytime nuisance was a man in a white cowboy hat and fringed deerskin shirt who called himself the "Rodeo Kid." Whoever was unfortunate enough to be on daytime duty often found himself attached-to by this supposed refugee from the Wild West. He would swing up his guitar and lay it on the reception desk, then ask to speak to the announcer.

The captivating Bea Brix, our beautiful blonde front desk receptionist, knowing full well the implication, would call the announcer in the control room, who would then emerge to see what this visit was all about. It was always about the same thing: the Rodeo Kid wanted a chance to strum his guitar and sing genu-wine cowboy songs on CKTB air. As there was no time slot free for such an impromptu performance, the announcer would have to say no. The Rodeo Kid would then ask to borrow ten dollars, as he was broke and had to eat.

The announcer, himself barely subsisting on his own paltry salary, would of course refuse. The Rodeo Kid would persist until the announcer finally had to dash back into the control room to change a record. There never was any explanation as to why the Rodeo Kid, phony or genuine cowboy, would choose a little eastern city like St. Catharines, Ontario, to make a bid for fame. He eventually faded from the scene.

A hopeful for a different kind of fame was the Reverend Raymond Brown. Rev. Brown had learned his trade under our local Billy Sunday, and had broken away to launch himself as a competing on-air messenger of God. He did have a program slot on the station, paid for by himself or his supporters, at nine o'clock on Sunday mornings. In those days, most radio stations were, by CBC regulation, off the air after midnight, coming back on at seven a.m. weekdays and Saturdays, but not until nine a.m. Sundays. For the Rev. Brown to get in his full half-hour of preaching the Gospel — with covert attempts at hints for money — the duty announcer had to be there a minute before nine to turn on the station and project Rev. Brown onto the air.

We three announcers, Norm Bailey, Ronnie Congden, and myself, resided on the station's third floor in two separate rooms. It was standard for us to sleep on Sundays until about five minutes to nine, and the duty announcer to dash down the two flights of stairs in pyjamas or dressing gown, and snap on all the switches to make the station airborne.

The Rev. Brown, in his rumpled old white suit and crushed pork-pie hat would be waiting impatiently outside the front entrance to be let in to commence his broadcast. As soon as the door was unlocked, he would sprint to the little studio and, with hat still on, champ at the bit until he was announced on-air. Then he would let fly for the next half-hour with his message. The Rev. Brown used no music. His mission was too important to be interrupted by song.

The Sunday morning came when my alarm clock failed. I awoke at about ten past nine. Norm was snoring in the next bed. I leaped up. It was then I could hear the faint sound of pounding on the front door three floors below. Rushing down the stairs, I knew it was the Rev. Brown. Arriving on the main floor, I could see him away down the corridor outside the front door, pounding with his fists on the glass as though Armageddon was upon us. I got the station on the air, then went to let the Rev. Brown in.

I don't remember the actual language the "reverend" person used, but it was not a benevolent good morning. He rushed into the little studio and announced himself when I switched him on-air. It sounded like a most animated sermon after that, but I turned down the control room speaker and got busy finding the next show's records.

A challenge similar to Ronnie's church remote, but potentially more difficult, was involved in the regular Saturday night hockey broadcasts from the St. Catharines arena by CKTB's colourful sportscaster, Rex Steimers. This was tougher because the arena, being sheathed in metal, stopped radio signals, leaving Rex unable to hear CKTB's on-air content by radio to have a clue as to when to start talking.

There was also no field-telephone link. But there was a time-honoured solution. Timing was the key.

The control announce-op prayed that all clocks were synchronized. I would read the commercial just before Rex's eight p.m. deadline, give the station break, then totally without assurance that it would all work, flip the switch to put Rex on-air. To avoid dead air, Rex had to start talking at exactly the right second after the operator threw the switch and instantly begin bellowing his hockey spiel into his mike. Miraculously, it worked every time, but only by hairline timing.

Rex Steimers was one of the more exotic figures in Canadian broadcasting. He was popular far beyond St. Catharines, wherever CKTB reached; he was well-known for his barking delivery of sports news and on-the-spot play-by-play coverage of all kinds of sports events. A rotund dynamo in his fifties, with red nose, tweed hat, red or plaid jacket, and hoarse sportscaster veteran's voice, he was larger than life, as many successful broadcasters were. Rex would arrive about half an hour before he was due to go on the air with his six forty p.m. sportscast, which he always ad-libbed around some wire service sports copy. Smoke from a cigar floating around him, he would instantly head up the grand staircase to have a drink — or two — with his friend E. T. Sandell.

At about six thirty-eight, he would descend in alcoholic glory, rip the latest sports summary off the teletype, and thrust himself into the small studio to do his sportscast — mainly from memory of his rounds of games, or visits with sports figures of the day. There were no portable tape recorders then to verify the faithfulness of his memory.

But there was no need to. Rex was simply entertainment himself. With a teletype summary of sports scores in hand, he would roar his way through the ten minutes of sports, hat clamped down hard, half-smoked cigar clenched in teeth, and then bustle from the station and disappear.

To be in the same studio with Rex across a narrow table with a mike in the center — as I often was — to introduce his sportscast, was a wake-up experience. As soon as I completed my commercialized intro and said, "And now, here is REX STEIMERS . . ." Steimers' vocal chords blasted out a "THANK YOU GIL . . . and NOW . . . last night the NHL . . ." loud enough to make you rear back in your chair, especially from the overwhelming wave of alcohol/cigar breath.

Rex actually worked in the daytime for the local Carling's brewery, the one taken over postwar by E. P. Taylor's Canadian Breweries Limited.

Beer and liquor advertising was forbidden on Ontario radio and in newspapers, so Rex was inhibited in promoting his employer's Red Cap Ale. But he always ended his sportscast with, "Have a good night, folks, AND . . . you can't beat a red cap for a nightcap!" He got away with it. I suspect he had the full approval of Mr. Sandell.

One of Norm Bailey's pals, when he worked briefly for CBL in Toronto, was an announcer named Bill Bessey. Bill, decades later, became a celebrated personality on CBC-TV with a western music and square-dancing program, on which he wore a big cowboy hat and the customary deerskin shirt. In 1943, however, he was a very thin, not particularly handsome, urbanized young man with a deep CBC voice. To see Bill, as urban a CBC announcer as you might find, as a TV cowboy star years later stretched one's credulity. But he pulled it off well, and his show was, for quite some time, a CBL-TV fixture.

Norm invited Bill over from Toronto for an overnight visit. For a day or two, Bill loafed pleasantly around the station with Norm, getting to know us all. Whether it was a job-speculating jaunt or just a social visit was never clear. In those days, to have a visit from a CBC-level announcer was a notable occasion. He even chatted on the air with Norm as a special guest.

What really cemented our brief friendship was the arrival by mail of a flat package a few days after Bill went back to Toronto. It was a phonograph recording off the air of the voices of several of us on-air types. Bill, it seemed, owned a do-it-yourself home disc-recording machine, a personal possession then about as rare as owning your own airplane would be now. This machine, actually a recording lathe, long predated tape recorders; in fact, it went back almost to the time of Thomas Edison.

In 1943, it was an intriguing novelty to hear your own voice on a record. Twenty-five years later, after having routinely recorded on tape hundreds of news reports, I'd have found it boring. But we played Bill Bessey's recording over and over until the disc wore out. There is no measuring radio announcers' egos.

A few weeks later, Norm and I rode the train to Toronto to visit Bill. CBL at the time was located in the Canadian National Carbon

Company building at Davenport Road and Bathurst Street. It was a strange experience going from CKTB in a stylish old mansion to a CBC facility in a rundown old factory building. And we didn't see Bill Bessey at all. He wasn't working that day and nobody answered his home phone. We headed back downtown, saw a show and had dinner, then hopped the train back to St. Catharines. I didn't see Bill Bessey again for another fifteen to twenty years, and that only on a small black-and-white picture tube.

So while listeners sat comfortably at home in their easy chairs with the radio on, probably reading their newspapers and enjoying a Red Cap, and the men having a quiet puff on their pipes, away off in a darkened CKTB control room a harassed announce-op might be struggling to maintain the illusion of sending a steady, untroubled flow of pleasing sounds over the airwaves and into home radio sets. To the trusting audience, he was doing what came naturally. To the on-air man (nowadays, often woman) there might be one more silver strand amid his falling hair.

there'd be some changes made

Until the early 1950s, when TV appeared, radio was generally regarded as a handy toy, a music box that could be left on all day and probably all evening, too. It was only partly listened to most of the time, but in the evenings when Jack Benny or other comedians were on the air millions were, as was said, "glued to their radios."

Most stations made enough money to satisfy their investors, pay their staff their modest salaries, and cover other costs. Radio, for the most part, was profitable — but not in the wild way it became. In its own low-key style, radio was changing society, though not noticeably.

The movies were doing most of that, in the eyes of those who bothered then about such things as social revolutions.

Radio was a good source of music of all kinds, even though it was of low-fi, mono quality. 1940s listeners weren't so critical of sound values. Many became music fans, and some even performers because of radio. You could pick up snippets of news from newscasts perhaps twice a day, based on newswire services such as Press Radio. But for "real news" you read the local newspaper.

There wasn't much rivalry between these three media. They were symbiotic; in fact, the newspapers were running free listings of radio programs — even with review columns of major shows; the movie theatres were paying for newspaper ads, thus helping to support the newspapers; and radio, still in the late 1940s, was not a significant rival for ad revenue with the newspapers. The movie theatres had no intermission commercials playing to captive paying audiences. The public took all three of them as alternate or simultaneous sources of entertainment and information, but they also read books and general magazines. The time of The Manipulators had not yet come.

Probably because before the mid-50s the Great Commuter Era hadn't yet begun either, the car radio was a little-regarded accessory, where it existed, far from being the boom box that some became in later times. What had been a dashboard decoration turned out to be a hidden secret weapon for advertisers, as city workforce commuters began to span far out on extended highways in the 1960s to reach their hinterland homes in the evening and head back to the workplace in the morning.

Next to coffee, radio became the commuter's morning waker-upper on the road in, and often the soul-soother on the way home. The ad agencies were quick to catch on. The daily commuter, tuning in for traffic reports, was a sitting duck for the ever-increasing shriek of radio ads. Trapped in a jam, he or she kept the radio turned up to catch traffic reports. The commercials popped in and had to be heard first. Every radio station began its own version of a morning and evening road show, a feature undeveloped before the commuter age. Out of reach of TV inside the car, it changed the character of radio back to a necessary daily popular tool despite the 50s cataclysm of television. Profits came back.

In the early 1940s, commercials were, almost without exception, one-minute blurbs politely read out by live announcers from written scripts. The recorded vocal jingle, so unnerving today, was rare. Local recording studios, if they existed at all, were found only in the biggest cities. Recording equipment for cutting discs was too expensive for small operations.

The record industry consisted of the giants: RCA Victor, Columbia, Decca, and Capitol Records, all turning out music, classical and swing, on 78 rpm records, and by 1960, on 33 1/3s. Some commercials for national brands did come on noiseless recordings such as our black glass discs, using the smooth, professional voices of big-time New York announcers as though they were live in our very own local studios. Even that changed as loquacious Canadian DJs began to displace them.

Before the arrival of tape-recording in the 1950s, artists performed in studios, while recording lathes cut tracks of their sound into grooves in metal discs. The newly-cut masters were then, in turn, used to press the final composition records in mass production for commercial sale. Recording of orchestras and other sound sources was a fixed operation confined to recording studios and labs.

Tape, on the other hand, suddenly made it possible in the late 50s to produce good-quality instant recordings with portable tape machines at almost any site, indoors or outdoors, at the drop of a hat, with no careful handling or processing necessary. Although studio recording with its controlled acoustics was still preferred, anybody could record jingles and commercials cheaply in any location with ease, and supply them to radio stations to be played on the air. Ad agencies again snapped to attention. The age of instant advertising, endlessly replayable, had at last arrived.

Not only instant ad recordings, but instant, on-the-spot, live news coverage came with the tape age too. Now newscasters not only read out news more frequently, but some became carriers of tape recorders and could leave the radio station to record and bring back the voices of local politicians, police, firefighters, and a vast variety of other news-makers. At first this facility wasn't exploited to the full, as station-bound news readers, rightly so, didn't regard themselves as news reporters. In addition, they didn't care much about leaving the comfort of the studio for some time-wasting, cold-weather news site. There had to be a major catastrophe or event to break out the tape recorder and head for the battlefield. Only newspaper reporters were fit for that kind of dogged news coverage.

By the mid-1960s, that was changing. Trained reporters began migrating from newspapers to radio, carrying bulky reel-to-reel tape machines and covering everything from local city councils to formal interviews with newsmakers on the scene. Now, on top of the new threat from television newscasts, the newspapers had to compete with born-again radio stations draining news sources with instant reports using actual voices, a process that ended the long-standing era of the newspaper "scoop."

Radio, even better than local TV, could scoop the newspapers every time, if the will were there. On radio, constant deadlines were possible, with reporters going live on the air at any time with news reports, while newspapers had set times for going to press. That, coupled with traffic reports for car-bound audiences, gave ad agencies incentive to pump up the radio commercial. Radio now prospered beyond the wildest fantasies ever held in earlier broadcasting executive suites.

More and more, radio stations became de facto extensions of ad agencies. In Ontario, lifting the ban on beer advertising on radio in the early 1960s sent a massive shot of electronic adrenaline into the veins of the broadcasting industry. The decision of the Ontario government to cut the advertising bonds of the beer industry was proclaimed by Opposition newspapers to be just a Tory accommodation for the giant Canadian Breweries Ltd. controlled by the Argus Corporation, a holding company in turn controlled by the Canadian industrialist E. P. Taylor and a few colleagues. This was, of course, instantly denied. Other provinces soon followed suit on beer advertising in newspapers as well as on radio.

Argus also happened to own Standard Radio, i.e. CFRB, Toronto, which by 1960 was my employer, and CJAD in Montreal. Besides us, Argus owned the giant Dominion Stores grocery chain; Domtar in Montreal; Canadian Industries, Limited (CIL); Canadian Breweries Ltd.; and sundry smaller companies.

As a CFRB staffer, I had the questionable honour to be present in the studio when Bill Deegan, the RB afternoon DJ, read out — mainly for the car audience — the first beer commercial ever to hit the airwaves in Canada. It was, of course, for the most popular beer of the time, Carling's, one of Canadian Breweries' chief labels. I watched and listened respectfully as Bill made advertising history, unaware of the rich impact that this new decision would ultimately have on radio, especially in sports shows.

Bill Deegan, incidentally, was the man who kept most of Metro Toronto and its surrounding area informed and entertained in the eight-hour power blackout of 1965, which put most of Eastern Canada and the entire Northeastern United States, including Toronto and New York City, in the dark. Bill stayed at the CFRB microphone the entire time, for many hours beyond his regular work schedule, while maintaining his usual affable, calm style, between records assuring folks that there was nothing to panic over, even though privately he really didn't know what was causing the blackout.

RB's emergency power system kept the station on the air throughout the crisis. It turned out later that a power relay at Niagara had blown and, in a chain reaction along the common hydro grid, blacked-out most of Eastern Canada and the Northeastern U.S. Deegan was later publicly honoured for devotion to the public interest during a time of possible panic.

Politics had no effect upon CFRB's broadcast content. In my 1960s eight-year career at RB, during which I did countless news reports from the Ontario Legislature at Queen's Park, Metro Toronto Council, and other political bodies, no one in top management ever suggested a change in my choice of what news to report or how to report it, even though I might sometimes feature the leader of the New Democratic Party or the Liberal Party in my reports over that "Tory" radio station, CFRB. The same was observed by the rest of the news staff. It was an honour system, faithfully followed.

Actually, I played it right down the middle in reporting Queen's Park political news, with no regard to whether the governing party was Conservative or not. In the case of the other regular newscasts, CFRB was truly Ontario's authoritative news voice. RB didn't have to worry about grinding a political axe for anybody. It already had an average audience of four hundred thousand *households* in a region with a population of about four million people, an incredible lead over its closest competitor, CHUM. Its middle-of-the road music-and-news format seemed to be just what the vast majority of Southern Ontario listeners of the time wanted. If a campaigning politician sought exposure on RB, he had to pay for it, or be treated equally with campaigners from all parties.

Across Canada, by the 1930s, radio was thriving. The major centres all had their distinctive broadcasting outlets. Only the paralyzing effect of the Great Depression and then the World War II freeze on new ventures put a halt to the spread of radio broadcasting further into Canada's major and minor centres. With the end of the war, the lid came off. Domestic goods were back in production.

Every large city began to acquire more stations, small at first, but quickly catching up to the older, established outlets. Competition, especially for the advertising market, skyrocketed. The sound of radio hit a higher key. That brought on the era of the radio "personality." Each station needed a louder, more attention-getting voice to stay in business. The advertising world made similar moves, adopting more and more the "advertising jingle" that jangled more and more nerves.

In the 30s and 40s, the "radio announcer" was a man of strong male vocal tones — the lighter resonance of women's voices were hardly ever heard — speaking only briefly between music recordings, or firmly authoritative in newscasts, delivering commercial "spots" in a clipped, impersonal style. The rule was formality of manner, no-nonsense, and with few (if any) provocative remarks, and certainly no bad language.

Somewhere in the late 1950s, the rules began to change. "Announcers" became "disc jockeys," then "DJs," then in later times "talk jocks." The concept of the sacred air vanished. Air personalities began to talk the lingo of the common people, or what they believed it was. Mostly it imitated showbiz, movie, and rock talk. As in all communications media, the old respect for the audience's sensitivities was being deliberately turned around into an assault on its standards. To provoke the radio audience meant to provoke some of its members into making phone calls, bland or belligerent, and expressing their OPINIONS!

Astute radio programmers saw a new opportunity — let the audience speak its mind! The Talk Show was born. And that's where it more or less stands in the new millennium. On the talk stations, opinion rules the airwaves, valid, voluminous, vacuous, or vicious.

The age of classical music broadcasting in private radio was fading as the post-World War II era moved in. The assumption was that people wanted "music to relax by," now that the tensions of war had subsided. Even the classical-oriented CBC radio eased up on "formal" broadcasting through its second Toronto English-language station, CJBC, founded in the 1940s. This, in fact, was the purpose of the new CJBC — to allow the public network anchor station, CBL, to continue higher culture with classical music and serious commentary, while CJBC introduced a lighter note in CBC music and chat shows. CJBC even went, for the CBC, to the extreme far end in music: i.e. *Jazz*!

First it was Dick Macdougal, who'd switched from CFRB with CJBC's founding, who "hosted" this departure. Dick ran a Saturday afternoon show, "At the Jazzband Ball" until he, sadly, died an early death in the late 1960s. He was followed by Byng Whitaker and Elwood Glover, who jointly took over the CJBC jazz show. Both, as longtime CBC announcers on CBL, had advocated a lighter approach for "The Corporation," and they finally got their chance to exercise it.

Byng — named after Lord Byng, governor-general of Canada in the 1920s, not Bing Crosby — held a respected reputation in lighter music and a children's musical program, "The Small Types Club." He discovered in the CBC music library a recording of a little-known children's song, "The Teddybears' Picnic," recorded in England on HMV by a very young English singer, Ann Stephens. Through Byng's frequent playing of the recording on his program, Ann gained wide recognition, at least in Canada, giving "The Teddybears' Picnic" a popularity that lasted for decades.

Elwood continued with the jazz program while Byng moved into other areas, never losing his attachment to jazz and popular music.

Somewhere in the 1970s era of the Trudeau government's campaign to spread the French language across Canada, CJBC became a unilingual French Toronto station, as it continues to be today. Elwood eventually switched to private station CKEY, Toronto.

A major revolution, of course, had already overthrown the music policies of private radio, led by a country boy named Elvis Presley. The proliferation of rock-and-roll groups became rampant, one after another capturing the audience then fading away themselves as new ones mushroomed. The Beatles broke the traditional North American view that British or European musicians had nothing worth exporting to this side of the Atlantic, and that nobody but Americans could dominate the music scene. No more were the Big Bands heard across the land, and Bach, Beethoven, and Brahms — not to mention Mozart — were relegated to infant FM radio.

CBL went on into the rest of the twentieth century as the CBC's pioneer cultural heavy, finally vanishing altogether from the AM airwaves in 2001 as CBC radio in Toronto went all-out FM, with CBC Radio One and CBC Radio Two. CJBC French also went FM. Thus CBC-AM, for sixty years the solid CBC radio anchor, died.

The radio heroes of the 1960s and early 1970s, and of the decades before that, have largely faded away. There were giants on the airwaves in those days, both in the public view and behind the scenes. It is a moot question which of the two varieties deserves the most credit for lifting radio programming from the feeble efforts of the 1920s, the desperate survival of the 1930s, the bolder 1940s, the re-grouping of the 1950s, the stability of the 1960s, the aggressiveness of the 1970s, the recasting of the 1980s, the formulizing of the 1990s, the straitjacketing of the 2000s, the — who knows what next? No doubt it was a normal evolutionary process, but there were powerful business minds driving it. They saw radio as a licence to make money.

The man behind the scenes, who for a time made the most business impact upon Canadian radio, as well as on newspapers, was Roy H. Thomson, later Lord Thomson of Fleet. He was typical — and the most successful — of the Canadian radio tycoons who pulled themselves up by their broadcasting bootstraps in the 1930s. Under him arose at least one other tycoon, Jack Kent Cooke. Between the two of them it seemed impossible *not* to make money on a colossal scale — initially from radio. In the late 1930s, they both came down from the North on separate chargers and swept up Radioland's potentials, running into hardly any brick walls.

Roy Thomson (1894-1976), famous for his bottle-bottom spectacles and jolly public persona, started from lowly beginnings in the 1920s. His father was a barber in Toronto. After getting some business training, he tried farming in the West, but gave that up to sell DeForest-Crosley radio receivers in Northern Ontario. Stories vary on how Thomson pushed things to start making his fortunes.

One version is that he started up a furniture moving company in Timmins and hooked his moving vans together with radio. In between radioing delivery orders to the truck drivers, it is said, he played recorded music over his radio frequency. Local people, having no other station to listen to, picked up the primitive Thomson broadcasting service on their radio sets for its entertainment value. For the astute Thomson mind it was a short jump to selling air time to local advertisers on the strength of reaching the local population when their guard was down. The Thomson communications engine was building up steam and getting underway.

Another, or additional, version of the Thomson success story has it that in the early 1930s, he picked up a radio licence held inactively by the Abitibi Paper Company under a dollar-a-year rental deal on a take-back basis. From a Toronto company he bought an unused radio transmitter for two hundred dollars plus some IOUs, set it up in North Bay, and put it on the air as radio Station CFCH. Still a radio set salesman, Thomson recognized that the discovery of gold near Timmins and Kirkland Lake meant that the Depression impact would be less there than in the rest of the province.

Soon he set up another radio station in Timmins, CKGB, in 1932, and then CJKL, in Kirkland Lake, in 1933. That's where Hamilton, Ontario-born Jack Kent Cooke — destined to become a Canadian radio multi-millionaire who eventually defected to California — first came into the Thomson camp. In 1937, Thomson made the former encyclopedia salesman manager of the Stratford, Ontario, station,

CJCS. Twenty-five years later, after a career as owner of the Toronto Broadcasting Company, which owned CKEY, Cooke left Canada for greener California pastures.

As the story goes, in the basement of the CKBG building in Timmins, there sat a printing press left behind by a newspaper, the *Timmins Citizen*, that went under when the Depression arrived. Thomson bought it for a song. That was the start of the successful *Timmins Press*. By 1935, Thomson had turned his new venture in newspapers into a daily paper.

Newspapers, unlike radio stations, didn't need government licences. So Thomson switched his talents to newspapers, but still kept his hand in radio. He and the publisher of the *Kingston Whig-Standard*, Rupert Davies, set up CHEX Peterborough in 1942, and CKWS Kingston in 1943. Eventually, in the television era of the 1960s, both stations evolved into TV broadcasters as well.

As time passed, Roy Thomson expanded his business interests into acquiring more radio stations and weekly newspapers throughout Ontario. His headlock on the weekly newspaper finally prompted the Ontario government to bring in legislation limiting the number of weeklies one individual could hold. Thomson had to accede, and sold off many of his weeklies and radio stations.

He went international, buying up newspapers, magazines, and printing companies around the world, from Florida to the U.K. He owned radio and television stations everywhere. He took up residence in Edinburgh and was later made First Baron Thomson of Fleet. In 1953, he turned over ownership of twenty-eight Canadian and eight U.S. newspapers to his son, Ken. Among his famous trophies were Scottish Television of Glasgow and *The Scotsman* newspaper.

Thomson capped all this with purchasing the Kemsley newspaper chain which owned the *Sunday Times* of London, and in 1966, bought the venerable *Times of London* itself. After the senior Thomson's death in 1976, Kenneth, the Second Baron Thomson of Fleet, concentrated much of his business interests in Canada, buying up such institutions as the Hudson's Bay Company and other ventures. He became known in Canada as just Ken Thomson, tycoon, art collector, and billionaire, one of the richest men in the world.

Thus came the apotheosis of radio as megabucks generator.

not enough cookes?

Meanwhile, Roy senior's one-time Stratford, Ontario station manager, Jack Kent Cooke (1912-1997), was busily building his own fortune through radio. In the mid-1940s, with very little radio competition in Toronto because of the domination of CFRB and the CBC's CBL, Cooke bravely launched into station ownership, buying CKCL (later CKEY) from the Dominion Battery Company.

In his career to that point, despite the business hangover of the 1930s Depression and the vagaries produced by wartime, Cooke was financially successful enough to buy a Toronto-based general interest magazine, *Liberty*, entering the world of "convergence" early, so avidly pursued later by huge communications companies. In 1951, he bought Toronto's Maple Leaf baseball team, affiliated with the AAA International League. Both are now long defunct.

Two decades later, after television came to Canada as the CBC's CBLT, Cooke's was one of nine media companies that made a bid for, at the time, the only available Toronto television licence. He and seven others lost. Almost instantly in 1960, no doubt seriously miffed by the licence award to the *Toronto Telegram* and not to him, Cooke pulled up his Canadian stakes and moved his Canadian-made millions to Los Angeles, where he amassed an even more fabulous fortune. His U.S. citizenship was signed by President Eisenhower.

Cooke, plunging deep into radio, television, and cable TV ownership, rapidly became a prominent U.S. broadcasting and sports industry figure. He'd already had a fling as a minor sports magnate with the Maple Leaf baseball team in Toronto. Five years after his arrival in L.A., and now owning or controlling several California radio and TV stations, he paid $5.2 million for the Los Angeles Lakers basketball team, ultimately acquiring players such as Magic Johnson.

Next came the National Hockey League expansion team, the L.A. Kings. In 1967, he built The Forum, a twelve million dollar stadium in L.A. for his teams to perform in. He promoted Muhammad Ali and Joe Frazier's first match at Madison Square Garden, calling it "The Fight of the Century."

In 1978, Cooke moved to Washington, D.C., where four years before he'd become majority owner of the Washington Redskins football team. A year later, he sold the Lakers and the Kings for $67.5 million in the biggest business deal in the history of sport. The same year, he divorced his first wife in a settlement of $49 million, which went into the Guinness Book of Records.

Meanwhile, the Redskins were apparently doing fine, with Cooke personally operating the club on a daily basis. In 1983, they won the prestigious Super Bowl (XVII) against Miami, 27-17; they followed that with wins in Super Bowl XXII in 1988 against Denver, 42-10; and Super Bowl XXVI in 1992 against Buffalo, 37-24. Yet, in 1987, Cooke said the Redskins were losing money, and announced he would build a new stadium, at a cost of one hundred and sixty million dollars, in

Landover, Maryland. The Redskins had already moved into a multi-million dollar training centre in Ashburn, Virginia.

Along the way, in 1985 Cooke had bought the *Los Angeles Daily News* from the Chicago Tribune Company. Horse racing being one of his grandest passions, naturally he owned a race track. Real estate being another sideline, he bought the famous old Chrysler Building, a landmark skyscraper in New York City. He collected antique cars, fine art, and racehorses. Despite his mania for business acquisitions, and having dropped out of high school early — like the typical Canadian radio personality — he was said to be well-read in literature and informed in politics, music, sports, and architecture.

Cooke was known, both in Canada and in his later U.S. career, as a tough, hard driver of business deals. People who worked for him in his radio, television, and cable TV enterprises seldom had a good word for him, in fact, some commenting in terms bordering on hatred. Others called him hard-working. No doubt remembering his start in life, just scraping by as a 1930s book salesman and on a salary of twenty-five dollars a week in his first managing job for Roy Thomson, he paid them a pittance for what was rather demanding and unusual work.

He forbade his Redskins players to attend the funeral of the previous team owner, Edward Bennett Williams, who was in Cooke's bad books. He is said to have disinherited one of his sons. One of his wives attempted suicide more than once. He had major marital problems, and one of his two sons died at fifty-eight. Cooke himself died of heart failure in 1997 at eighty-five, at his plush home in northwest Washington.

He will probably never be listed in a Canadian radio hall of fame, having eschewed his native country and applied his staggering business talents elsewhere. He is not well-remembered by those in Canada's radio world who came close to him. Who knows what he might have done for Canadian broadcasting if he'd won the TV licence competed so intensely for in Toronto in 1960?

His continuing memorial is the Jack Kent Cooke Foundation, which he set up to provide funds for promising students for higher education — funds which he didn't acquire himself for his own education. But after all, you might ask, did Jack Kent Cooke really need it?

While radio was constantly expanding everywhere from east to west in the 1920s and 30s, numerous entrepreneurs were rising from among the ranks of pioneering on-air and non-on-air radio people, many still active and influential into the 1980s, and even the 90s. Private radio, despite public radio CBC being intertwined through its early powers of governance, has always been one of the more aggressive industries in the world of private enterprise.

The word "competition" isn't strong enough to describe the interplay between radio stations and companies owning a multiplicity of radio, and later television, stations. In many cities, rivalries between local stations have been murderous, equal only to the wars between newspapers. The struggle for advertising revenues and popularity ratings began early in the industry's history. While the competition was always savage, it can be understood because of the fact that private radio's only lifeblood is advertising. Minus commercials, obnoxious and brain-numbing though some may be, radio and TV would be dead in the water.

It has been said that the strongest, and probably final, bastion of purely *private* enterprise is radio/TV. The development of both media has been based on the momentum of driving individual ambition and determination to dominate. The examples of Roy Thomson and Jack Kent Cooke testify to that. But there were others who achieved great success in their own careers, and with the enterprises they headed for a time. Few other communications media have pivoted so much on the personalities and drive to win of the individuals who ran their stations on massive ego-overdrive.

In the 1980s and 90s, victory over the other guy was often achieved by buying out and taking him over, as practiced by Big Business everywhere. Earlier, in the 30s and 40s, the battle was mostly a clashing of swords between Station A and Station B, with Station C possibly getting in a few thrusts as well. Eventually, one might have its back forced to the wall and give in to the more ruthless swordsman with a sell-out.

This kind of duel was common until the 1970s, when "corporate takeovers" came into vogue on a large scale in the world of private radio. Then, some seasoned radio people metamorphosed into broadcasting tycoons and became corporate raiders, seeking out targets to acquire and remodel into components of new radio empires. In the process, favourite old local stations — many beloved by the public — were sucked up, and had their program formats changed to fit some new industry trend or formula.

Typical was CFRB in Toronto, the trail-blazing Canadian pioneer founded by Ted Rogers, Sr. and his brother Elsworth, in 1928. Ted, of course, was the inventor of the power vacuum tube, which freed up home radio receivers from the storage battery and introduced sets that could just be plugged into electrical wall outlets, revolutionizing the radio-sales world. Over its next fifty-seven years, CFRB held the lead as Canada's most successful private radio station, putting on the air diverse personalities whose names became household words. Its format of "good music," news, and commentary made it the top station by far in Canada's most lucrative market. Generations grew up with CFRB in their ears. CFRB was "always there," solid and dependable.

In 1985, it became just another station in the string owned by Slaight Communications, which bought out a number of Canadian broadcasting companies, including Standard Broadcasting, Ltd., the owner of CFRB and CJAD, Montreal. Allan Slaight, radio entrepreneur *par excellence*, had been the proprietor of CHUM, Toronto. Founded in 1946 by a group of war veterans, CHUM grew from a minor station into CFRB's main competitor in the 1960s, though at the time it couldn't hold a candle to CFRB's popularity among a mature audience.

For most of that era, it was a mature audience that held family purse strings, so they were, therefore, the preferred ad target for sophisticated commercials. That began to change after about 1970, when youth culture power began to take hold in entertainment. CFRB's longtime good music and news format was ignominiously scrapped, and it became a talk-radio station, abandoning music entirely. Thus was told an ironic tale of a mouse swallowing an elephant and outgrowing the elephant.

o pioneers!

In the earliest days of Canadian broadcasting, not many colourful tycoons of radio were around yet. Most stations were started by local enthusiasts in the grip of the new medium's technical ethos and techniques, fascinated by radio's power to project voices and music far beyond the room they sat in, relishing the effect all of this had on a still-naive radio audience.

At first they had freedom to say over the air whatever suited them, knowing their chatter would command an audience far greater than could be packed into a local hall or wherever they might otherwise be able to speak to a listening group. The experience was similar to the power felt by today's users of the World Wide Web talking to the universe in chat rooms. All that changed with the entry of government control into radio broadcasting.

The early radiophiles were, for the most part, tinkerers with only amateur status in electronic knowledge. Enthusiasts who weren't busy operating ten-watt stations were most likely building their own radio receivers at home, first of all crystal sets, then the vacuum tube version with headphones and multiple plug-ins for a group to listen together at the same time. Loudspeakers were still to come. Whole families, ears clamped in headphones, sat around in a group, still and dead silent, to hear radio's fascinating sounds on their single radio set.

A stranger from outer space walking into an early 1920s living room might have been astounded to see adults and children with odd apparatuses on their heads, immobile and wordless as ancient Greek statues. Perhaps the spell would break with a group laugh at something only they could hear. As a family activity, listening to early radio in this fashion was hardly a binding force, but it did have the primitive elements of joining a group together to enjoy a common experience, even though that would require a jolly reconstruction around the dinner table of what they'd all heard. Some genius put a set of earphones in a large bowl or metal box and amplified the tiny sound into something loud enough for several people to hear at once. That was just a stop-gap for a time. As parts became more available at the neighbourhood radio store, the tinkerers added loudspeakers that really broke the silence.

All of this, in the early 1920s — even into the 30s — helped maintain a keen interest and dedication by self-taught radio amateurs, whose tinkerings made it possible for their families and friends to hear the marvelous tinny sounds of hockey broadcasts, comedians' voices, and the music of the radio stars of the times. Sports, of course, stirring the fans' imaginations with the excited tones of the sports commentator and background crowd noise, dominated many evenings around the radio. The Joe Louis boxing matches, where Joe usually laid low his opponent early in the first round, kept the fans enthralled. It was an era of excitement and discovery, of reaching into the little known, of a kind of ethereal magic. Soon, though, the magic would begin to fade, as advancing technology brought the ethos of radio down to earth. Radio became practical.

By 1926, the loudspeaker reigned supreme. Magnavox produced a big-horned speaker that really did bring the voices of radio into the home for all to hear together. Thus the era began of the all-pervasive audio that would permeate the air, blossoming later into the age of the boom box, the Walkman (back to the earphones), and the inescapable overhead serenadings of shoppers in malls and grocery stores.

The stampede to louder and higher-quality home audio didn't really begin until the 1950s. This followed on the heels of the Long Play microgroove record, introduced by Columbia Records with a super-bang in 1948. The debut of the 33 1/3 rpm LP was a scoop by Columbia that, at first, sent the recording and radio industries reeling — not to mention the home music enthusiast, who then had to invest in new equipment to play the LPs. Columbia pulled off its coup after a secret three-year project recording on LPs some of its most popular contract artists such as the Benny Goodman Orchestra and numerous musical aggregations.

These recording sessions were held secretly until the new system was released to the market, and Columbia had its backlog of LP recordings all ready to SELL! When they announced the new Goldmark system, they caught their rivals completely by surprise. The secret had been tight. RCA Victor and Capitol Records scrambled to catch up. Even though Columbia offered to share the secrets of the LP, RCA hurriedly brought out its own little 45 rpm disc with the big hole in the centre and the special player to play it on, hoping it would counteract the LP. By that time, Columbia's release of its new system and its backlog of recordings got to the public so much earlier, that the LP was already its darling. The 45 had limited success, but the 33 1/3 LP was far ahead. Ultimately, RCA and Capitol, plus new companies, took up Columbia's offer to share, and produced their own 33 1/3 LPs.

The record revolution also doomed the decades-old original 78 rpm disc. It was probably time to scrap the old 78 players anyway. They'd served their purpose, evolving from the early phonograph with its giant metal horn through the cabinet-encased gramophone to portable turntables with tone arms holding cartridges for steel- or diamond-tipped needles. But the 78 hung on until its final scrapping in the 1980s. It actually had shown remarkable survival power in the face of the LP and the tape recorder. Then, with the passage of time, even those recording miracles bit the dust in the age of the compact disc.

For radio, these upheavals in sound reproduction were expensive. With each one they had to replace their "modern" music equipment with expensive new systems in order to play the torrents of recent recordings. The revolution that began with the magic of tape recording following World War II segued into the age of FM stereo, then forced the stations to upgrade and eventually move full-time into the world of "reality audio" in the 1960s.

Stereo was not all that new. Electric and Music Industries (EMI) in Britain launched a stereo system in 1933, using 33 1/3 standard-groove discs. It didn't catch on. In 1958, Audio Fidelity in the U.S. and the Pye and Decca companies in the U.K. put a stereo record on the market. At the same time, commercial radio was starting to adopt better sound equipment and move into FM stereo broadcasting. At last radio was able to do justice to the high quality of sound already contained in records.

The ultimate crisis — so far — arrived in the form of the Compact Disc (CD). Once more all older recording and playing equipment was swept aside. For dealers in electronic home playing equipment this has been an enormous windfall, bringing customers into stores looking for the latest way to enjoy music. The technical revolution goes far beyond that: making it possible to record on CDs at home even the most recent music, by way of the Internet. This, as might be expected, set off a severe reaction by recording companies and, by default, recording artists, all of whom rightfully expected royalties as payment for their musical efforts. Technology had superseded integrity.

World-wide, legislatures triggered laws to protect the creators of music media from music fans' piracy and the illicit profiting by fly-by-night — and even foreign conglomerate — recording upstarts. Obviously, if such piracy went on ad infinitum, the music creators might no more find a point in producing recordings just for the pirates to profit by, and might retire from the scene. At this date, that doesn't seem to have deterred the thieves nor, so far, the artists. Regulations of a certain kind have been brought in, but whether they will permanently foil the wholesale theft of music remains to be seen.

Once the power given by advertising became recognized as a viable way to put radio on a commercial basis and thereby expand into being an indispensable link with the rest of the world, radio entrepreneurship started its booming career in earnest. The blip of World War II over with, the struggle for audience rating superiority intensified. New stations blossomed in cities where there had been only one, stirring the older managements to find new methods of undercutting the upstart opposition. Advertising agencies sprang up everywhere. In the big centres they might be branches of long-established U.S. companies who'd already honed the techniques of audience manipulation to a fine edge. But Canadian agencies quickly emulated. Radio and radio advertising were becoming almost one, with advertising on top.

The way also opened for the more daring to invade new fields of opportunity. Across the country there were companies, often newspaper organizations, that had been already getting into radio. As an example, in 1928 the Winnipeg/Brandon, Manitoba, family firm of Sir Clifford Sifton acquired the *Regina Leader* newspaper, the founder in 1922 of CKCK Regina. The newspaper eventually became the *Leader-Post*. Sir Clifford was a cabinet minister in the late nineteenth and early twentieth century Canadian government of Prime Minister Wilfrid Laurier. In 1953, Clifford, Jr., a son of Sir Clifford, took over the *Leader-Post* and the *Saskatoon Star-Phoenix*, plus CKCK and CKRC, Winnipeg.

Clifford Sifton, Jr., graduated from Osgoode Hall Law School in Toronto in 1915 and joined the Royal Canadian Artillery as a junior officer, later rising to major. He was involved in most World War I Canadian actions in France, was wounded three times, and got the Distinguished Service Order (DSO) for gallantry.

Sifton became a shareholder in Wentworth Radio Broadcasting Ltd., formed in 1938 by the Calgary company of Taylor, Pearson, and Carson. The company owned radio station CKOC, founded in 1922, in Hamilton, Ontario. Under Sifton, in 1953, CKOC entered an equal partnership with Hamilton's other radio station, CHML, and the *Hamilton Spectator* newspaper to found CHCH-TV, at the time one of the only two television stations to be created in Ontario outside of the CBC's CBLT-TV, Toronto.

In 1960, Armadale Communications, which was founded by Sifton, took over full ownership of CKOC, and sold off its CHCH-TV holding. Getting out of TV in 1960? Odd decision. Well-known in Canada's radio world, Clifford Sifton sat on the board of trustees of the Canadian Association of Broadcasters (CAB) as a strong advocate of broadcasting freedoms. He died in 1976. Armadale Communications dropped out of broadcasting in 1997 after divesting itself of its radio and TV interests.

Other influential Canadian communication chains, like Southam, Inc., added broadcasting to its holdings on top of its newspaper owner-ships. Its flagship paper, the *Hamilton Spectator*, where the Southam fam-ily began their newspaper career in 1844, as noted above, became a part-ner in the new Hamilton TV station, CHCH-TV in 1953. The instiga-tor of this coming-together of communications giants was Kenneth D. Soble (1911-1966), founder and MC of Canadian radio's first amateur show, "The Ken Soble Amateur Hour," in the early 1930s. He'd begun in Toronto radio as a sixteen year old in 1927. Starting the amateur hour when he was twenty, he set up a Toronto production company and went into the business of producing radio shows.

Senator Hardy of Brockville bought CHML Hamilton in 1935, and hired Soble as manager. The station prospered, but by 1944 it was up for sale again. Ken Soble was one of three who made a bid for the station, and he won. The other two were Roy Thomson and Jack Kent Cooke. Under Soble's ownership of Maple Leaf Broadcasting, CHML continued to prosper as a keen competitor with CKOC and even, for listeners west of the Toronto area, with Toronto stations as well.

As a teenager in 1940, I was an avid fan of Norm Marshall's "Club Clambake," a daily CHML program devoted to the then-booming Swing Era of the Big Bands. I learned a good deal about swing and jazz from that show, and it paid off for me later when I went to work for CKTB, St. Catharines, where I ran jazz and swing programs. Norm, incidentally, began his career at CKTB, too. He went on to become a well-known news- and sportscaster on CHCH-TV.

Ken Soble was a visionary of broadcasting. His partnership with Sifton and Southam proved to be a powerful lever to pry loose one of the three TV licences the CBC had assigned to Toronto. Between Soble and Kenneth Pollock of the Kitchener, Ontario, firm of Electrohome, a manufacturer of radios and also the owner of Kitchener's CKCO Radio, they convinced the governing CBC in 1953 that Toronto should not have a monopoly on all available TV licences but that the two remaining licences should be awarded in cities well outside it.

So were founded in 1954 CHCH-TV, Hamilton, and CKCO-TV, Kitchener, for a time the only Southern Ontario competitors of CBLT-TV, the CBC's pioneer television station in Toronto. Despite their success, the two new television stations had to be CBC affiliates. The persistent Soble, however, managed to achieve independence from the CBC. CHCH-TV thus became Canada's first independent television station.

Ken Soble wasn't finished with his vision for Canadian radio and TV broadcasting. In the early 1960s, he showed an unusual level of foresight in proposing to the new Board of Broadcast Governors, set up by the federal government to take broadcasting regulation out of the hands of the CBC, that a TV network be created, using the brand-new technology of earth satellite linkage to relay programs to ninety-seven private stations across the country.

With satellites only just beginning to be used for bouncing communication signals around the world, Soble's concept seemed far-fetched to some. Soble died aged fifty-five, and his space-age Canadian radio/TV network idea died with him. That is, until it came about anyway at a later date. Since then, relaying every kind of electronic signal by satellite is commonplace for not only broadcasting networks but for every user of telephones around the world.

the beat goes on...

The above expression encapsulates the exact mindset behind "rock radio" that blossomed in the 1950s. New stations popped up on the dial playing the current Top Ten hits dictated by a rigid formula. The long-established "good music" stations took a back seat with the youngsters. The "disc jockey," with his raucous shouting style and chatter, and commercials that became more and more hysterical, pounded out a continuous rock-and-roll drumbeat around the clock, wasting not a second of time. The new format seemed to have been just what restless youth wanted, even later as adults, at least according to the record promoters.

One might ask whether the frantic formula of 50s and 60s rock radio was a portent of a growing frenetic pace of life, or whether it was the actual instigator of such a pace amid the youth of the time. It then mushroomed into more and more of the "lifestyle" that became general as the "pioneering" youth grew older, still responding to "the beat." Having acquired a fixation on a fast, steady pace to live by, were the baby boomers and their successors themselves responsible for today's overdeveloped sense of urgency — and with it frustration — that characterized life in the 70s, 80s, 90s, and now in the new millennium?

Tiny portable radios and cassette players, complete with headphones, became a most prized birthday and Christmas gift by 1970. Earplugs were available for the older set. Now teenagers could go everywhere with the throbbing cacophony of rock music and rock DJ screech continuously pouring into their ears, supposedly shutting out the world of adult values and its demands for order and conformity.

Such exposure, near to being electronic brainwashing, was bound to prime the juvenile mind for a fast pace of their later lives, all in the name of ad-induced demand for the products created by demand itself, possibly followed by the frustration of personal non-achievement. Today, the beat still goes on, just as the old rock-and-roll slogan had it, but an aging audience dances to it.

Just as the LP with its microgrooves destroyed the ten-inch 78 rpm disc, the mainstay of the recording industry for half of the twentieth century, the Compact Disc roared in somewhere in the 1970s like an unholy storm. All of a sudden, the 78, the LP, the tape reel, the tape cassette, and the sophisticated equipment to play them on were all swept aside. A new miracle of sound recording, the stereophonic CD couldn't be matched for fidelity and durability. Radio stations scrapped their old audio-playing gear, music stores rebuilt their counters to hold the smaller CDs, and the electronics industry marketed their new CD players with enthusiasm and verve.

Once again, the public went for the novelty of the new, not necessarily because the old was not workable anymore, but because the CD epitomized the new age. Now here was something more that folks hadn't known they should have. The old recordings, with their fabricated aura of mustiness, the symbol of an age gone by, were consigned to the dustbin by the new technology. Rapidly, "78," "LP," and "record player" became quaint terms that would bring puzzlement to the faces of young music store clerks. The question leaps to mind: From whence comes the next new technology that will blot out the CD?

Regardless of high tech, the music goes on. Mozart sounded great even on the old twelve-inch 78s, but sounded better and better with every technological wave. Recording companies had orchestras regularly make new recordings of the great classical composers, creating new demand simply with new interpretations by new conductors. There always were classical orchestras. With their presence, the world of the CD boomed with big help from the upspringings of rock and country music.

Radio, from earliest times on, might have remained a utility much like the telephone if it hadn't been for recorded music. Recorded music "made" radio. In the 1920s, radio was a thing to be held in awe, emitting sounds by some mysterious means, a force from beyond the ken of the average person. The voices that came into one's living room from *wherever* must be, some guessed, those of spirits from the ether and not people you would call up on the phone like you would Aunt Effie in Tillsonburg.

By the 1930s that superstition had evaporated, but even then you wouldn't likely call a radio station any more than call city hall to complain about noise. Both kinds of efforts would be lost in the great shuffle. Radio was considered really for listening to music you didn't have in your own collection of 78s, not for listening to extended talk from disembodied voices.

Of course, the more hardheaded generations that emerged after World War II saw radio for what it was: a medium based on electricity manipulated by ordinary earthlings. Calling up a radio station on the phone was just a thing you might do if something you heard bothered you. That eventually grew into the phone-in show. Some thinker at some radio station saw value in putting callers on the air with their beefs and opinions. Maybe somebody out there agreed with them. Or maybe they disagreed. A controversial opinion begot a few more calls, and then some more, and lo, you knew that people were listening! Great for audience ratings!

Like almost every other innovation in the entertainment world, new roles for radio came mainly from the U.S. — New York City or Los Angeles. In Canada, the trend toward talk radio in the 1980s seems to have moved in all at once in the major cities and in some of the minor ones. At last the person in the street, if such exists, could berate the world with his/her staunch opinions. Radio was now becoming a father confessor — not of sins, but of beefs and disguised resentments against life. As that developed, music became something in the way of "free expression." Many stations dumped it. For them, the concept of radio as a cultural medium was for the birds. Culture by radio rapidly faded into the ether.

the biggest daddy
of them all

A major case in point was the venerable Ontario institution of CFRB Radio, Toronto. Founded by Ted and Elsworth Rogers in 1927 as an incentive for people to buy their Rogers Majestic radios, CFRB rapidly became a pillar of Canadian radio broadcasting. The launch of CFRB with its first studio in the foyer of the one-time Massey family mansion, later Ryan's Art Gallery, and still later Julie's Restaurant, at Jarvis and Isabella Streets, Toronto, was an important event for all Canadian radio of the time.

The opening on February 19, 1927 was gala. The Rogers' saw to it that the Toronto upper crust would be there, complete in tuxedos and white ties, and dazzling gowns — with probably the odd rebellious flapper in a short skirt sprinkled about. The launch had its moment in the next day's papers, despite their aversion to radio as a competitor. It was a Toronto social event of the new 1927 year.

A three and a half hour radio show, a first for Canada, featured leading local theatre stars in an invisible — as radio essentially is — variety show hosted by Jack Arthur, the era's impresario of Toronto main events. The show went on until midnight. For those tuned in on their radios, CFRB's superior sound was the hit of the evening. Nobody had ever heard such clarity as the new station's signal. Ted Rogers had put all his technical skill into his brainchild's on-air debut.

To the working staff of two it was even more important. Wes McKnight was the pioneer station's only announcer and Bill Baker was the operator, CFRB's first employees. They didn't know it, but they were both being launched on careers that would enshrine them as Canadian radio trailblazers, Wes as sportscaster supreme and Bill as the technical genius of the leading decades of Canada's most successful radio station. CFRB grew in both broadcast coverage and popularity. Even people in the sub-Arctic and out at sea could tune in to CFRB through its short-wave station, CFRX.

Ted and Elsworth Rogers were ready to try anything in broadcasting that came along. This even included the chance of starting the first Canadian television station. They applied to Ottawa in 1938 for a licence to run one. Sadly, World War II got in the way. Long before that, foreseeing that the Ryan location wouldn't be adequate for their future broadcasting ambitions, the Rogers brothers went big-time by opening an entirely new station at an uptown Toronto location, 37 Bloor Street West. It moved there in 1931, and stayed until 1965, when Procter and Gamble erected a sleek new building at St. Clair Avenue and Yonge Street. CFRB took over the entire fourth floor.

With its coveted first dial frequency of 1030 kc (kilocycles, the precursor to megahertz) and a power of one thousand watts, CFRB had a clear, static-free radio channel that reached most parts of Southern Ontario in the days when only a few stations existed in other cities, and there were only about one hundred and thirty-five thousand radio sets in the whole country. By 1930, there were eight commercial stations in Toronto. Only CFRB survived them all. The others had call-letters now long-forgotten: CKCL, CKGW, CFCA (the *Toronto Star*), CKNC, CPRY, CNRT, and CNRX.

By the early 1930s, the Rogers station was broadcasting organized programs of news, sports, and music. One of its regular sportscasters was Foster Hewitt. By 1935, now at the maximum transmitting power of ten thousand watts permitted by the CBC, CFRB had become the most commercially successful Canadian private radio station with the largest audience in Canada, remaining so for decades to come. Pre-eminent in music, news, and sportscasting, it could not be matched in popularity by any other Canadian private station. It was the envy of radio enterprises from coast to coast and a psychological crutch, in its own way, for thousands who apparently had no desire to listen to any other radio station.

The Rogers brothers were legendary pioneers in North American radio innovation and manufacturing. Ted's invention of the power vacuum tube in the mid-1920s put an end to the then-conventional battery-powered radio sets. Their leaky storage batteries' sulphuric acid was rapidly removing the finishes on thousands of fine polished buffets and tables across the continent, much to the fury of thousands of housewives.

Ted Rogers' brainchild allowed radios to be plugged in to any electrical wall outlet, rendering obsolete all existing sets. Ted's invention of the power vacuum tube, using regular household alternating current, was finally taken up by all makers of radios. Radio was at last freed of the stigma of furniture destroyer. Sales boomed.

Other, less powerful stations appeared, but devoted fans stayed tuned to 860 on the dial, RB's second frequency (it was later, by CBC decree, moved to 1010). One listener who said she loved the CFRB music-and-news format told me her dial setting never shifted from RB, wherever it was. CFRB, she said, was like Eaton's department store: always there. (Alas, when the new millennium came in 2000, Eaton's itself was no longer there and CFRB was an all-talk station.)

The first Canadian broadcasts of the National Hockey League went out over CFRB air on the voice of Foster Hewitt. The rousing recorded call of a posthorn and the animated vocal delivery of Jim Hunter, newsman extraordinaire, hauled in huge audiences at six-thirty every evening and brought farmers racing in from their fields to hear the latest gripping news à la Jim Hunter, a revered institution all in himself.

With a radio station of enough broadcasting power to push a clear signal into most parts of Southern Ontario, and even up into the northern holiday cottage country of Muskoka and Haliburton, CFRB was hard to beat. As time and the Second World War came and passed, other stars bombarded the airwaves from CFRB's studios. John Collingwood Reade began his nightly commentary on the war in the 1940s; Gordon Sinclair, the maverick *Toronto Daily Star* reporter, started his brassy newscasts in 1947 after his fifth firing from the *Star*; John Bradshaw got people up in the early hours with his country-style music show and farm news. The same year, Wally Crouter took over the morning slot from John and rose to being a radio icon for thousands of risers, while John continued his rural style in a mid-morning show.

"The Crout," as Wally was known to his friends, came from CHEX, Peterborough, to CFRB in 1946 after Canadian Army wartime service, and stayed for fifty years. During those five decades, he became a legend in Canadian radio as the epitome of the morning man. He seemed impervious to wear and tear despite turning up every morning at five a.m., often arriving straight from the most recent party. Fast-talking, full of wit, and always bantering with his control operator, Bev Edwards, who became almost equally as famous though never being heard on the air, Wally was popular and formidable on the golf course and in the social whirl as he was as the top early morning broadcaster.

Of champion calibre on the golf course himself, he played with the best: Arnold Palmer, Bob Hope, Bing Crosby and Perry Como, to name a few, all notable fairways stars themselves. The question among fellow CFRB staffers was "When does this guy ever sleep?" He was always wide awake at five in the morning, and for hours after.

radio's chairman
of the board

A growing rumble — like approaching thunder — was coming down the corridor. You knew enough to get out of the way fast. Not to do so might interrupt a dash from the newsroom by one of CFRB's top money-makers to address roughly half a million folks, starting in about ten seconds. The thunder was from the hurrying feet of Jack Dennett, morning and evening CFRB newscaster supreme, dean of Canadian private radio newscasters, the voice of reason and stability as much as his older colleague, Gordon Sinclair, was the voice of dissembling disturbance.

Jack was heading for the studio to perform one of his twice-a-day newscasts. Without his smooth baritone delivery of the latest news, lives in a multitude of homes where CFRB was regarded as the only station on the dial apparently couldn't go on. As he burst into the announcing studio, the *tah-tah-te-**daaahh*** of a recorded posthorn would be sounding over the air and all through the halls and studios of CFRB.

That was the call that alerted everyone that "The News With Jack Dennett" was about to begin. This was the same sound that had long ago even brought farmers in from their fields to glue their ears to every word of Jack's broadcast predecessor, Jim Hunter, CFRB's dean of news from the 1930s to the late 1940s. Now it cleared the decks for Jack Dennett.

Jack was anxiety in the flesh. At ten seconds to eight in the morning and ten seconds to six-thirty in the evening, with the news script that he'd been preparing for the previous two hours clutched in one hand, he would do his dash after one last desperate look at the newswire teletypes for any late bulletin and then slide before the microphone just as an announcer blurted out, "Now, here's Jack Dennett!!" As calm of tone as his anxiety was high came the reassuring Dennett voice, "Good morning/evening, everyone . . ." and the solid pronouncements on the latest state of the world would begin.

It wasn't mike fright. Jack was a perfectionist, with all the uptightness that goes with that personality type. Not one second of time could be wasted once Jack began to prepare his newscast. He used every shred of it, even to that last attack on the Teletype machines. Jack knew perfectly well that I, or one of the other newsmen back in the newsroom, would rush any important late-breaking bulletins into the studio to him if he were already on the air. But he carried on with his anxious habit of waiting until those last few seconds before heeding the call of the posthorn.

It wasn't as though he feared for his survival. As one of the top CFRB stars, he was totally secure in his job. Besides that, he was also well-known as a regular between-periods pitchman for Esso on the "Hockey Night In Canada" CBC broadcasts of National Hockey League games. On top of even that, he and a real estate partner had made a fortune in a late-1950s housing development in the town of Agincourt in Scarborough on the eastern edge of Toronto. His own house was built on Dennett Drive, named, of course, after himself.

This affluence helped reinforce his image as the chairman-of-the-board conservative who distributed the dividends of the day's news to thousands of listener shareholders. He was certainly a living reassurance of the "straight" and traditional way of life and could be caustically acid on the air about "weirdos and freaks" who were disrupting that life and trying to destroy its values.

Like most stars of broadcasting, he was fiercely proud of his huge listening audience, but at the same time deathly afraid of somehow losing it. There was about as much chance of that happening as there was of Soviet Chairman Khrushchev kissing U.S. President Kennedy during the infamous Cuban Missile Crisis. It told on him physically. When he was going about tearing up swaths of Teletype copy as he prepared his newscast, he would grimace and complain of stomach upset, a sure sign of an anxiety attack. It all could well have had a bearing on his final passing at the age of fifty-eight.

Still, he could take with a straight face the pranks his workmates played on him from time to time, such as the trick with the water jug. Some would recall him as naive and an easy mark for practical jokes they might shoot at him before they finally broke up and gagged with laughter over having "had" radio's great voice of authority. It was a tradition around RB to take down any overly serious authority figure, as Wes McKnight — the picture of unsmiling solemnity — would have attested. Wes was burdened with the job of General Manager after a long career as one of CFRB's most memorable, and original, voices of sports broadcasting. It seemed to weigh heavily on him, and like Dennett, he was also a target of the station's jokesters.

Tall, stoutish, balding, rapid-fire of tongue and quick of movement, Jack Dennett had been a handsome man in his younger days. Well aware of his vulnerability to practical jokes, he had his own way of striking back. Though his victimization by such practiced jokers as Jack Dawson and Wally Crouter was legend around the station, he usually got even. Crouter frequently referred on-air to Jack, approximately his own age, as "my father" or "the old fella" who would arrive in a moment to do his eight a.m. newscast. He might also recount outrageous and imaginary anecdotes about Jack's parsimoniousness and his tight grasp of a dollar, itself not entirely fictional.

Actually, while busy polishing his news copy back in his office next to the newsroom, Jack couldn't hear any of these comments. But through others he was aware of Crout's loving campaign of friendly on-air character assassination. On one occasion, Jack broke his dedicated last-second arrival in the studio and arrived early as an uninvited "guest" on Crouter's show, which immediately preceded and followed his eight a.m. newscast.

He boldly entered the studio where Wally was talking on air. He took over the mike and congratulated Wally on his sixtieth birthday, and marvelled for the radio audience at Crout's continuing and amazing sexual virility at that advanced age. Crouter was actually around forty at the time. That might account for an acquaintance of mine who regularly listened to the Crouter morning show actually believing that Wally was a "jolly old fellow" who was on the verge of senility.

Dennett was also well-known for his inability to take a real vacation. After much pressuring by his colleagues to go south for a few weeks to get his nerves settled down to a reasonable level of turmoil, Jack might finally agree that he did indeed need a rest. A substitute newscaster would be scheduled to fill his news slots and away he'd go on his vacation. But barely six days would pass before he was back in town. He would call off the substitution and next morning come bustling in at his usual arrival time of six a.m., tearing into the teletypes with renewed vigour.

In spite of what might seem his class-conscious outlook, Jack followed no such distinctions in his relations with any of the CFRB staff, ranging from our eighty-year-old office boy, Milt Rothermel, to Winks Cran, the president. He knew everybody by their first names and carried on running battles of mild insults mingled with daily greetings, especially with the control operators. These worthies were traditionally at feigned odds with the on-air people in deadpanned slurs and poker-faced contempt, sometimes in dead seriousness, but usually in jest. Over the intercom between studio and control room, Jack could match the bored operators with the same kinds of put-downs.

He also was the featured newscaster on CFRB's FM station, CKFM, with a half-hour newscast that ran every weekday evening at seven p.m., to a different audience, almost immediately after he finished his six-thirty AM newscast. Therefore, after his arrival at the station, between five-thirty and six-thirty p.m., he was always getting two different newscasts ready. But he couldn't possibly put together a half-hour newscast in the twenty minutes between the two. So between five-thirty and six-thirty he would write a few minutes' worth of his FM newscast, then rush to the nearby FM studio where an operator waited to record him, a bit at a time, on tape.

Then he would rush back to his office to work on his AM newscast. Bit by bit, through a series of these mad dashes he would have most of the seven p.m. FM show on tape before six-thirty, when he would then dash into the AM studio to do his six-thirty newscast. After that was completed, he would dash back to the newsroom to rip another ream of news copy off the teletypes to get the latest, latest news and hurry to the FM recording studio to complete the seven p.m. tape.

With only minutes or seconds to spare, the operator would reel Jack's half-hour tape back to its start, and roll it exactly at seven o'clock for the FM audience, the Dennett voice all through it carrying the very sound of calm control of the situation. Jack would then leave, and listen to himself on his car radio tuned to CKFM as he drove home to Dennett Drive. No wonder the anxiety.

The last time I saw Jack Dennett was in 1972, three years before his death, when he responded to my invitation as McMaster University's then-Assistant Director of Public Information to come over to Hamilton by our special helicopter shuttle for Toronto news people to witness the opening of McMaster's giant new Health Sciences Centre, one of Ontario's biggest buildings at that time. The opening ceremonies of this $80 million medical marvel were done by Ontario Premier William Davis, with an accompanying range of dignitaries from Ottawa and the general health field. Jack Dennett was among them.

With forty thousand visitors touring the building on opening day, it was an event of Jack Dennett stature. Jack, pleased with his helicopter ride, was every inch his old smiling, affluent self as we walked around the new facility. Afterwards, he 'coptered back to Toronto in time to do his evening newscasts. This time, though, he must have had less anxiety than usual, as he filled most of his AM and FM newscasts with glowing reports to his half-million listeners of the new McMaster Health Sciences Centre. Jack Dennett's dignified stamp of approval was the final official touch to a momentous inauguration.

the elite of the airwaves

With the passage of time and the possibility of a CFRB television station by 1960, the roster of RB stars grew. The attractive and coolly personable Betty Kennedy was hired from the *Ottawa Citizen* newspaper, which she'd joined as a reporter at sixteen, to do a mid-afternoon interview show, hosting a parade of public figures from Hollywood movie stars to political giants. Subsequently well-known across Canada to the audience of the CBC-TV show "Front Page Challenge," she did her actual daily work on her one-hour CFRB program, "The Betty Kennedy Show" over twenty-seven years. Active on corporate boards, government advisory committees, and educational bodies, she ultimately authored two books, *Hurricane Hazel* and *Gerard*, and was named to the Canadian Senate in 2000.

At CFRB, Betty was a calm and self-possessed "pet" of the station, yet without personal involvements. Devoted to her husband, Gerard, she projected the same persona on the air, coming over as someone you could confide in and receive consideration from, if appropriate, or encouragement, if needed. Her interviewing technique was to ask brief but penetrating questions and then let the interview subject have the floor. This was effective, and different from the interviewer who wants to talk as much as or more than the interview subject. Audiences responded by tuning in to her show every weekday in the hundreds of thousands.

Bob Hesketh, at first a freelance relief man for Gord Sinclair and later full-time staffer, did his popular, wry "The Way I See It" daily commentaries on life in general. Bob, a former *Telegram* sports columnist, was the man waiting in the wings to replace Sinclair when he retired. Sinc was just turning sixty-five at the time.

But being no ordinary working man, he put off retirement for years, staying on and on for two more decades, never retiring, and passing away at the age of eighty-four. In the meantime, Hesketh did his commentary, and finally took over from the late Sinc.

CFRB, by 1960, already had an elite cadre of on-air personnel, each with his own following: Wally Crouter from early to mid-morning; Jack Dawson following (later succeeded by Earl Warren when Jack became station manager); Bill McVean in the early afternoon before Betty Kennedy took over until four; and Bill Deegan with his rush hour "Rolling Home Show" into early evening.

Literally soaring high above them all in the CFRB traffic helicopter was Eddie Luther, next to being an airborne god to those jammed in traffic down below. A Second World War Royal Canadian Air Force ex-pilot, Eddie was the original Toronto radio pioneer eye-in-the-sky, reporting on Metro Toronto's traffic tie-ups from the helicopter every weekday morning and evening rush hour, and the holiday weekend road race. Wes McKnight, the first RB announcer, major sportscaster, and by then station manager, remembered hiring Eddie postwar — still in his RCAF uniform — and putting him right to work on the air.

Eddie worked for years as a disc jockey and newscaster, but in 1961 the station leased a chopper and a pilot from Pegasus Airways, who kept it in a barn at E. P. Taylor's Windfields Farm on the northeast edge of Metro Toronto when it wasn't up in the air. Eddie was launched in it daily from there to report on Metro Toronto traffic jams. The service was an instant hit with the prisoners of traffic.

Proof of Eddie's huge following among evening rush hour drivers (and the number of car radios tuned to CFRB) was demonstrated when Eddie, as dusk began to fall, would remark on the air from the 'copter that it was time for everyone on the roads to turn on their car lights. Metro Toronto main arteries and Highway 401 would light up instantly in a long blaze of headlights as listening drivers reacted. The people power of CFRB was confirmed.

The chopper was useful to RB in several ways. As the pioneer in traffic helicoptering in Ontario radio, possibly in all of Canada, for CFRB there was great public visibility in the air as well as audibility on the air. When the yellow helicopter was hovering over Highway 401, listeners knew that CFRB was keeping a kindly watch on affairs below. It enlarged the image of the benevolent broadcaster.

At times I did relief trips for an absent Eddie. There were almost no restrictions on where we could fly around Metro Toronto. We could do a great circle around Metro, sitting the chopper down wherever an accident report from the ground was advisable, or picking up TV film, photographers' pictures, or other items at a waterfront pad for delivery to the Toronto international airport for sending to out of town media by air. Slipping into and out of the airport was quick and easy, until the Department of Transport ruled it off limits, for whatever reason.

Rules became tighter about altitudes and no-fly areas as time passed, and also when another station decided to go into helicopter traffic reporting, too. But while it lasted, it was a great novelty to coast low over skyscrapers, spotting cars like multicoloured bugs scooting around down there among them and viewing the bare construction sites of future landmarks like Yorkdale Plaza and the Toronto-Dominion Centre, just getting under way. The older downtown Toronto was still intact. There was no CN Tower yet, nor the seventy-two-storey First Canadian Place, nor its shorter mates of the 1970s.

We had a request from the Metro Police to borrow the helicopter so that two plainclothes policemen could look for the body of a boy who'd drowned in the Humber River and was believed to have been carried downriver to Lake Ontario. To keep a CFRB presence with them in the 'copter, I was sent along. We spent the afternoon slowly sliding sideways low down along the Toronto waterfront and then later virtually skimming the surface of Grenadier Pond, while the two cops scanned the waters below. They had no success in spotting the body. It was startling, however, to see in the pond giant goldfish over two feet long just under the surface. They'd never be visible from shore. The big carp were the grown-up result of people dumping their unwanted pet goldfish in the pond to get rid of them.

Throughout the evenings, there was a parade of specialty CFRB programs of a kind now long-extinct in Canadian radio: Ray Sonin with his "Calling All Britons" show that especially enraptured British immigrants and longtime British-descended Canadians with its Old Country themes; Ruby Ramsay Rouse, the old-time organist, who carried on for a devoted audience the romantic music of earlier times; John Collingwood Reade, with his daily analysis of the news; and Walter Kanitz, with his collection of European music recordings; Alan Small did the evening classical music programs both on CFRB-AM and on CFRB-FM, one of the earliest Toronto FM stations; Wayne Van Exan played music through the night until the return of Wally Crouter at five a.m.

Another station, CHFI Toronto, was the first commercial broadcaster in Canada to send out a signal strictly in FM stereo. Oddly, one of the partners in the new station launched in the early 1960s was Ted Rogers Junior, son of Ted Rogers, father of the power vacuum tube. Ted Senior died suddenly in 1939 at the young age of thirty-nine. Joel Aldred, a former CBC announcer-turned-successful U.S. TV network car-company commercial pitchman, was a partner in CHFI with Ted Junior, with the backing of the Eaton's department store family.

They set up the station in the vacuum of FM radio absence, and for a time the station was a hit. Ted Rogers Junior went on to found and head Rogers Communications, giant of Canadian cable TV, which also bought up Maclean Hunter Publishing, ultimately acquiring nineteen radio stations across Canada.

The remarkable tenure of CFRB as Canada's most successful private commercial radio station — though not the oldest — is literally the story of private broadcasting in Canada. Certainly it was located in the most active commercial region in the country, but so were other stations, competing vigorously for the same audience.

Successful radio has always been successful because of its on-air personalities. CFRB held that unique and lofty position for over fifty years of some of the most dynamic times in Canadian life. Time, unfortunately, has its way of erasing favoured institutions. Those fifty years of incomparable human achievement are now part of Canadian history. Their like will not come again.

a world up in the air

This is where I, for a time, must come back into the picture. At the risk of accusation of taking an ego trip, nevertheless as a one-time CFRB staffer I can tell from firsthand experience *some* of the story of the world of Canadian radio up to the 1970s. There may also be others across the country who have their own memories of life in various Canadian radio stations, but these are largely unrecorded.

Having in 1946 abandoned interest in working in radio again once World War II was over and the army was through with me, I became a newspaperman with no intention of ever returning to the broadcasting field. When my return to radio unexpectedly happened in 1960, I'd been working for the previous ten years as a reporter and a sub-editor at the *Toronto Daily Star*, having been projected directly into a career with the *Star* upon graduation from the University of Western Ontario's School of Journalism in 1950.

With a background of commercial printing, radio, some freelance magazine work, and being a war veteran and university graduate at twenty-five, I must have seemed a reasonable choice by Canada's biggest and most successful newspaper. Whether that was really so or not, I couldn't foresee that I'd eventually also be invited to join Canada's biggest and most successful *radio* station as assistant news director. But the times between and after were, if nothing else, tough.

Although I was only mildly aware of it, in early 1960 CFRB and eight other broadcasting companies were just getting ready to battle over the only television broadcasting licence remaining available to a private Toronto broadcaster after the CBC launched Canada's first TV station, CBLT Toronto, in 1952. The award would be made by the Board of Broadcast Governors (BBG) — the federal body succeeding the CBC as Canada's broadcasting overlord — to the best presenter of a licensing proposal in a grand week-long meeting in March of 1960.

The BBG had previously decreed that allotting licences for television be done very sparingly to deserving companies in Canada's major cities, usually one per city. Because of its big population, Toronto was assigned three. This instantly inflamed a demand from smaller cities outside Toronto that only one go to Toronto, and the other two be awarded where there was capability for TV services. As mentioned, the attack was led successfully by Ken Soble of CHML Radio, Hamilton, and Ken Pollock of the Electrohome electronics manufacturing company and CKCO Radio, Kitchener.

The nine TV-seeking companies also included Standard Radio (CFRB); Spencer Caldwell Productions; Consolidated Frybrook Industries, Limited, under Jack Kent Cooke; Granada television of Great Britain; Marconi Broadcasting of Montreal; the *Toronto Daily Star*; the *Toronto Telegram* newspaper, allied with newly-formed Baton Aldred Rogers Broadcasting; and Maclean-Hunter Publishing. All were armed to the teeth with arguments for granting them the TV licence. The BBG's hearings in the big Oak Room in Toronto's Union

Station were a week of dazzling presentations by the nine companies, on film and in person. Most stars of Canada's communications world were there — if not involved in the hearing, then as observers.

The Baton Aldred Rogers Broadcasting company was a formidable alliance of some powerful figures in Canadian business. John Bassett, owner of the *Toronto Telegram*, was chairman of the board, owning forty per cent of the stock. Joel Aldred, wealthy Canadian and well-known New York commercial TV announcer, was president. The Eaton family, of department store fame, was also a key backer of the company.

Aldred and Ted Rogers, Jr. — son of CFRB's founder — owned thirty percent of the company. Paul Nathanson of Sovereign Films owned ten percent, hockey broadcaster Foster Hewitt ten percent, and another ten per cent was owned by Heathcourt Boulevard Investments, made up of leading Toronto lawyers Eddie Goodman and Charles L. Dubin (later a Chief Justice of Ontario), who were receiving shares in lieu of legal fees for setting up the new company. Bassett told the hearing that the board had agreed to vest control of the company in the *Telegram*. Within eighteen months, this array of corporate broadcasting talent would be fiercely divided and broken up, with "the *Tely*" taking all.

CFRB's early application in 1938 for a TV licence seemed to make Standard Radio a safe bet in many media observers' eyes. Back in 1938, television in North America was still a pipe dream. It had been proven practical in the United Kingdom, through the mechanical system invented by Glasgow's John Logie Baird, since its first successful broadcast in 1928. In North America things were a little slower. Ted Rogers, Sr., was alert to the Baird phenomenon, and in the true Rogers style made a bid to be the first on the continent to send pictures through the air. Alas, World War II was on the horizon, and no development of TV was allowed.

When the opportunity came twenty-two years later, CFRB got ready to go on the attack. After all, it seemed to make sense that the best-established and best-respected radio broadcasting company should win the right to form the first private commercial television station in Toronto. In Ted Rogers' day, there'd been no competition. CFRB's decades-long career in live radio broadcasting — just a step away from

TV — was, in the opinion of most who knew the field, the logical recipient of the licence. The odds were for RB.

The BBG had for months been holding hearings across the country, starting in Vancouver, granting TV licences in the major cities. Hadn't Prime Minister John Diefenbaker declared that no television licence would be granted to any newspaper? That narrowed the field to real broadcasters, in the view of CFRB management, and RB was the leader in the field. No account was taken of politics.

At least a year before the BBG hearings were to begin, the CFRB group went to work preparing a presentation that would knock all the competition out of the ballpark. Under W. C. Thornton "Winks" Cran, Standard Radio president, $2,134,129 was placed in a bank by Standard, earmarked for setting up the new TV station. There was also a bank credit of one million dollars if needed, and surplus CFRB earnings would be available as well, bringing the pot to about five million. Not many Canadian companies of any kind were worth that much at the time.

New stars-to-be, such as Betty Kennedy, were hired from other operations. CFRB sent programmers, producers, technicians, and fledgling cameramen to over forty television stations in the U.S., U.K., Germany, Italy, and Canada to learn TV studio techniques as well. While RB always had outstanding newscasting people, such as Jack Dennett and Gordon Sinclair, it had never had a specific news department, with news specialists covering actual events, writing and taping their stories, and going on the air with specially written original on-site news coverage.

Radio news had usually been, with the exception of Dennett and Sinclair, a matter of a general announcer ripping the latest Canadian Press news summary off the teletype and/or sitting down at a typewriter and "scalping" the latest newspaper edition for the top news. This would not be good enough to impress the BBG. News would be a major part of a TV station's programming and would have to carry the authenticity of true journalistic coverage to satisfy the BBG. For that, a whole news staff would have to be hired.

This, apparently, was where I came in. For some time, as a *Toronto Star* staffer and as part of my editing-desk work, I'd been assigned to phone in top news to CFRB as soon as the first *Star* edition came up from the pressroom, and as the later editions arrived. To me, it was a chore added to my responsibilities on the fast-moving editing desk. Probably because I'd mentioned some time in the past to our martinet chief "slot man," Willis Entwistle, that I'd once worked in radio, he farmed the job on to me to call in the news to RB. It didn't really take an ex-radio man to do this, but I got the job anyway.

The arrangement had been made between CFRB and the *Star* some time that year to exchange news. There were old links between RB and the *Star* despite the Conservative Party leanings of the former and the Liberal Party leanings of the latter. Gordon Sinclair, Foster Hewitt, Jim Hunter, and others were originally *Star* men. The news exchange worked one way only. CFRB almost never acquired news before the *Star* did, not having a professional newsroom to do original coverage.

A minor move, it was probably a preliminary to creating a real CFRB-TV newsroom to meet the BBG's standards. Bill Hutton, an editor with the Canadian Press Toronto office, and who'd once been a TV anchor in New Brunswick, was appointed news director. Bill was the man who, after I'd been feeding news to RB for a month or two, asked me in February if I'd like to join the RB staff. Feeling I'd had my day as a *Star* man, I accepted.

The arrangement eventually was extended to include the services of the *Star*'s outstanding and now-legendary police reporter, Jocko Thomas, to phone in regular news from Police Headquarters, making him a beloved fixture in the CFRB radio area. Jocko became another Hutton hire, and we were workmates again.

into the den of the lion

When Hutton led me into the office of Wes McKnight, CFRB's station general manager and sports broadcaster emeritus, for a final interview, I couldn't help recalling the wintry day in early 1944 when, as an eighteen-year-old CKTB announcer, I'd journeyed by train from St. Catharines to Toronto to attack the barricade of Big Radio and join the announcing staff of the eminent CFRB.

On that day, I'd trudged all the way up Yonge Street from Union Station in bitterly cold winter weather, climbed the ugly stairway to the second floor of 37 Bloor Street West, and asked the receptionist at the front desk to speak to someone about a job. The man who emerged from the labyrinth of corridors was, I now knew, Wes McKnight himself, then at the peak of his fame in radio and the world of sports. It took him about two minutes to glance me over and grunt that there were no openings at that time. He didn't take the time to say, as Ken Soble had earlier at CHML Hamilton, that if I wasn't called up for military service to come back. The rejection was final.

Now, in the warm light of 1960, the story was different. He didn't recognize me, of course, but I recognized him. I felt like a star football prospect when he rumbled from behind his desk in his well-known resonating radio tones: "We're interested in you, Gil. When can you start?" Naturally, I said *right away*. The cobwebs of 1944 were swept aside. With no effort I'd become a CFRB man at last.

Even in 1959 it was known that the day of the big award was coming soon. By the turn of the decade, CFRB's preparations for the BBG presentation became more fevered. The station's management was on red alert in every aspect. Winks Cran was the engine driving all his executives to their utmost efforts to ensure that CFRB would win the licence, himself greatly respected — if not feared.

Much was riding on it. Cran, a former executive of Decca Records, Inc., of Great Britain, was a dynamic Scot with a cultured English accent who'd already put CFRB well into the black as a prolific profit-maker. Cran explained his nickname was hung on him in childhood because of his love of the "Wee Willie Winkie" nursery tales. His business methods were hardly typical of the nursery, however.

Jack Dawson, manager under Cran, found him rather tyrannical. Once, when Jack was interviewing staff one by one in his office about our new contracts, I read mine over, agreeing with all the terms but one: vacation time. I pointed out to Jack that, although the contract seemed to give me two weeks vacation a year, I actually got only one week, plus compensating time for statutory holidays, all of which I worked except for Christmas Day. I'd been given the choice of Christmas or New Year's Day as time off.

To Jack I suggested I should have another week's holidays to bring my vacation time to an actual two weeks, plus holidays. It seemed reasonable to me. So, pen poised, I was about to write in the change. Jack said quietly, and somewhat ominously, "I wouldn't, if I were you." I instantly felt the presence of Winks Cran hovering over us, although he wasn't actually there. I got Jack's meaning. I signed the contract as it was.

Around the same time, I was heading home late one night after finishing up my newsroom evening stint when I spotted a familiar figure leaning against a building just outside the Bloor St. subway station. I recognized a CFRB executive and assumed he was waiting for a taxi to take him the rest of the way home. Then I saw he was far beyond even the point of total drunkenness. Concerned about his self-responsibility, I suggested I get a cab for him. He roused himself enough to shun the idea in unmistakable, though slurred, words.

After spending a few minutes more with him, I gave up. As I went into the subway station myself, the last I saw of him he was feeling his way along the building wall in the direction of Yonge St. I concluded that he was on his way to find a hotel, if he could. I reflected that being an executive at CFRB was not a happy state to be in.

Standard Radio, owner of CFRB, bought from the Rogers family in 1947 by E. P. Taylor's Argus Corporation, though still the private Canadian broadcasting leader through the fifties, didn't really pick up full steam in money-making until Cran arrived in the top executive suite. Oddly, though surely a total coincidence, the word "*cran*" was a World War I French army term for nerve, or just plain guts.

Le cran was the word favoured by the French officer corps to describe bravery in the field. Where the odd French officer was deemed to lack "*cran*," he was instantly downgraded and replaced. Winks Cran certainly had his share of *cran*, or its civilian version. He drove his RB troops hard, spreading the concept of *cran* everywhere. Still, he was highly respected among the staff.

There was rehearsal after rehearsal in the old CFRB theatre studio, supervised by a film producer, honing each participant's part in the coming presentation to a fine edge. The result was a half-hour colour film of the stars who would be featured on CFRB-TV: The beautiful Betty Kennedy, doing "Woman's World"; popular early morning man Wally Crouter; famous globe-trotting newsman Gordon Sinclair and authoritative-voiced Jack Dennett; top sportscaster Wes McKnight; and most of the well-known RB voices, in the flesh.

Sinclair could already claim he was Canada's first TV personality, as he'd been televised to, and in front of, crowds in a closed-circuit TV system at the Canadian National Exhibition in 1939, long before there was any chance of a commercial Canadian TV station. Most agreed at the 1960 CFRB viewing session that the CFRB film presenation for the licence was a "sock-o" production. It would knock out the opposition with the greatest of ease.

I was on board at CFRB by the beginning of March 1960. The station was still at 37 Bloor Street West, occupying the entire second floor of a rambling two-storey building built in 1928 that was originally meant to last for twenty years, making it obsolete in 1948. But here it was 1960 and the building and CFRB were still there. Of course, that was to change as soon as the TV licence was obtained. Then there would be a new building, the latest in TV and radio technology and design.

The tension of waiting for the granting of the licence could be felt in the air. As more progress was made on the presentation, executives charged around with greater vigour, exuding more and more confidence. Old hands like Reg Wilson, a control room operator for the previous twenty years, shook his head in disbelief, unconvinced that any good would come of all this "fooferaw."

Jack Dawson, program director and chief announcer, would come into the cubbyhole that served as a newsroom, giving a great mock-conspiratorial wink at Bill Hutton, as though to say the licence was "sewn up." There was advance jubilation all around. Winks Cran, tall and handsome, strode in and out of the newsroom beaming. Then the Big Day arrived.

Each day for over a week the Board of Broadcast Governors would file into Union Station's Oak Room, itself theatrically half-darkened with lighting directed at the board and the presenter. They took their places at a long table up front before a sea of seats full of concerned media trade attendees in the dark areas, prepared to listen to all arguments. The chairman, Andrew Stewart, would call on the next scheduled presenter and then all of the board would sit silently as the most influential figures of private showbiz management took the stand and extolled the fantastic superiorities of the companies they fronted.

There were lawyers of the first water; tycoons like Spence Caldwell and Jack Kent Cooke, already famous, or notorious; contract stage and radio performers, all of whom couldn't say enough good about their clients, companies, or sponsors; and, in the back rows, small fish such as myself and junior CFRB staff people, listening obediently. Every one of the presentations was cleverly, even spectacularly, carried out.

At CFRB's turn, Winks Cran rose to present his station's considerable claim to run a television station in Canada's biggest commercial market. He went over strong points like RB's original request for a TV licence in 1938, the long experience in broadcasting live talent, its great standing in the Toronto area and beyond for broadcasting personalities and good programming, and its huge listening audience.

The newspapers, perhaps being applicants for the TV licence themselves, were not kind to the CFRB presentation. While all week they'd been giving breathlessly detailed accounts of the hearing, when RB's turn came on the Saturday, they gave very little space to its case. On Monday, the *Globe and Mail* did a roundup of the nine applications, including tidbits like "the CFRB women's editor" (Betty Kennedy, who was not so designated) who would run a show instructing girls on how to trap a husband. When BBG counsel Graeme Haig asked about this, Cran, in his cool British style, quipped, "I suppose that would come under the heading of Sports." Great laughter broke out.

Thereafter came a cavalcade of the more famous RBers, all of their familiar voices booming confidently over the PA system in their famous tones, sending to the audience a warm assurance that CFRB was the one-and-only choice one might make for the new era of Toronto TV. The CFRB film, in colour although TV was still all black and white, produced as though an actual television broadcast, news and all, was shown as the clincher.

Standard Radio's lawyer, Joseph Sedgwick, brother of a once well-known CFRB president and longtime head of the Canadian Association of Broadcasters (CAB), showed a list of one hundred and seven letters from Toronto and district municipal officials, organizations, and individuals supporting CFRB's application, and read out parts of them. After some other comments, he said CFRB stood alone in its application, holding no commitments with any organization.

The *Globe and Mail* reported Baton Aldred Rogers also had a list, including many of the names on CFRB's. When *Globe* reporters checked with some of the religious leaders on the RB list, they said their support was for the *Telegram* and any other endorsements were invalid.

Then, enter predictable Canadian suspicions of the Power of Big Business, with a whiff of party politics drifting in. Graeme Haig noted that 49.5 percent of Standard Radio's stock was held by the Argus Corporation, headed by E. P. Taylor.

"Who owns Argus?" he asked. A heavyweight member of the Argus board, M. Wallace McCutcheon, answered, "Argus is owned by 400,165 Canadians."

the truth will out

"Does somebody hold controlling interest?" Haig continued. McCutcheon said "a group holds thirty percent of the Argus stock." Haig pursued, "Who is in the group?" McCutcheon would name names if the board thought it was relevant. The board *did* think it was relevant, said Chairman Stewart. McCutcheon replied that the controlling group was E. P. Taylor, Colonel W. E. Phillips, John A. McDougald, and himself. He didn't mention the other members of the Argus board, J. H. Radcliffe, J. Elsworth Rogers, Samuel Rogers, Major General A. B. Matthews, and Edward W. Bickle.

"Then it is conceivable," Haig suggested, "that if this group controls Argus, in turn it could control Standard Radio?" "Yes," McCutcheon answered. There was silence as the realization of this chilling thought — in the minds of some — sank in. Standard Radio's presentation then moved on.

The *Telegram* was also turning the screws on CFRB through its news columns. It headed a story with "Insist CFRB tell of E. P. Taylor Links." The very mention of E. P. Taylor hushed the whole proceedings and the audience, too. Canada's leading and most successful industrialist, once Liberal Prime Minister Mackenzie King's Minister of Wartime Production though a dedicated Conservative, Taylor, survivor of a mid-Atlantic mid-war torpedoing, had been responsible for the production and successful delivery of wartime supplies to Canada's armed forces. Postwar, Taylor brought together a diverse group of smaller companies into one overall organization, creating work for thousands of Canadians early in the 1950s recovery from the war, rustled up old ingrown suspicions of such a man's motives in wanting a TV licence. And besides, though he was a Conservative, he wasn't any pal of Prime Minister John Diefenbaker, reigning in Ottawa at the time.

Well, among Argus's member companies were Canadian Breweries Limited, which had corralled most of the smaller brewing companies of the land. And here we were being told that E. P. Taylor was asking for a TV licence! Could that mean he would then twist the Ontario government's arm to allow beer advertising on CFRB-TV? Advertise *beer*? In *Ontario*? Of course, other TV stations — or radio stations — would hardly oppose that lucrative advertising prospect if they had any say in any of it, but the idea would sink like a stone in other quarters. Mrs. Leslie Frost, wife of the Premier of Ontario at that point, was a staunch member of the Women's Christian Temperance Union (WCTU), a group who'd lobbied before and during World War I for prohibition, with its later disastrous results. However, on top of this, in even other quarters, Taylor, McCutcheon, Phillips, et al. were well-known as Tories in a day when the Ontario government was totally true-blue Tory. Ergo . . .? The Great Canadian Compromise seemed inevitable.

The BBG wasn't ready to pronounce its verdict for several more weeks. That was a time of eager tension in the CFRB Bloor Street domain. The question of E. P. Taylor had no relevance for us. Rather, how long would it be before we'd all be in a sparkling new building with a television studio complex added to a state-of-the-art radio broadcasting operation? How different it would be from working in the flimsy old rabbit warren of a Bloor Street building, dim and gloomy and with an assortment of garbage cans and contents gracing the back entrance that most of us had to use. A new millennium of broadcasting would begin!

The big news hit with the impact of a truckload of rocks. The Toronto TV licence, the Board of Broadcast Governors decreed, would go to the *Toronto Telegram*, owned by a single entrepreneur, John Bassett, who presumably hadn't thought about beer advertising. In spite of Prime Minister John Diefenbaker's previous declaration that no TV licence would go to a newspaper, the BBG's decision was just the opposite. A lone newspaper, owned by a proprietor with no background in broadcasting of any kind, was to be the king of private Toronto television, with only the CBC as competition. The earth under Canadian broadcasting's feet must have trembled. It certainly seemed to at CFRB.

I was sitting typing a news story when, out from the corridor to the newsroom the tall figure of W. C. Thornton Cran rocketed past. My glimpse of his face showed a countenance dark pink as a thundercloud, eyes flashing behind his spectacles, and a firm purposefulness in his walk as he strode toward Bill Hutton's little glassed-in office cubicle as though he were going to impale Hutton on the spot as the cause of the failure. But he didn't. I could see Cran slumped in a chair across from Hutton behind his desk. I could hear nothing. I didn't need to. Winks Cran was crushed.

Long face after long face paraded through the newsroom to Hutton's office in the days following, as though Hutton either had a good shoulder to cry on or could prescribe some kind of solace to soothe the catastrophe. Jack Dawson was devastated. Wes McKnight didn't even appear. Bitter words flew about from Jack Dennett as he prepared his next newscast. Gordon Sinclair predicted that the new TV station wouldn't last more than a year under John Bassett, and would be a load of headaches for everybody involved. "Who needs it?" he bellowed. Sinclair, of course, for decades a *Star* man, had no hopes of appearing on a TV station owned by the *Toronto Telegram*, the *Star*'s arch-rival.

Men and women from all CFRB's parts seemed to see the newsroom as the place for mourning. For weeks, anyone who came in had something to say, sad or bitter, about the BBG decision. The bitterness went beyond RB's perimeters. Jack Kent Cooke, his fury at being rejected almost too much for him, announced he would get the hell out of Canada and take his millions with him. Which he did, and became a billionaire in California. Spence Caldwell, stealing words from Roy Thomson, called the award "a licence to make money" and went about his business.

In the *Telegram*'s triumphant full-front page trumpeting of its victory, John Bassett announced that the new station would be called CFTO. He and the *Telegram* appeared to be the only happy campers in the Toronto media world. E. P. Taylor and his colleagues were silent. The *Star*, which had been one of the nine applicants, didn't make much of the development. Beyond that, as the dust settled, the public waited eagerly for the new CFTO to go on the air. The staff of CFRB awaited the axe.

beyond the blue horizon

Oddly, there was no break in stride by the CFRB on-air folk. After the decision of the BBG was given fair play in the newscasts, the topic of television licences vanished from the air. Only at the grand opening of CFTO, the new TV station, was the subject mentioned again, then dropped. No more was there a buzz of executive activity passing through the newsroom. Life went on, but at a more subdued tempo. The era of pure radio returned. No axe fell. CFRB was still Toronto's dependable old friend.

Gord Sinclair counteracted his misfortune of not becoming a CFRB-TV star by getting himself on the CBC-TV's new panel show,

"Front Page Challenge," thus launching into new fame. Betty Kennedy also became famous countrywide as a panelist on the same show. They both continued to hold huge audiences on CFRB, however, numbering at peak hours up to four hundred thousand households, in a Toronto area population at that time of about three million people.

Contrary to what might have been expected, the Argus Corporation was not, apparently, mad at us for not getting the TV licence. About May, personal printed invitations went to every member of the CFRB staff, and probably CJAD's too, to come to a reception at the historic renovated farmhouse of J. "Bud" MacDougald, just across the road from E. P. Taylor's Windfields Farm.

Both Windfields and the farm of Bud MacDougald, an Argus board member, were leftover countryside lands well outside Toronto acquired after World War II by Argus. They were to be part of the eventual further development of the Don Mills suburban development launched by Argus as a "planned new city" concept on the northeast fringe of Toronto in the 1950s.

These farms were lands unused in the development but with possible future use in Don Mills expansion. Windfields, of course, became E. P. Taylor's famous horse-breeding farm, which produced the sensational Northern Dancer, winner of the Triple Crown of the racing world.

Though it was a black tie event and we male staff rented tuxedos from Syd Silver's shop on the ground floor of 37 Bloor West, Madeline and I didn't really expect anything grandiose. We drove up to the front porch entrance of the MacDougald place in our rusty seven-year-old Ford. I was startled when a red-jacketed car jock popped up at my window with greetings and deftly opened my door. Catching on, we slid out of the car, I gave the keys to the jock, and he drove the old wreck off to some obscure parking place, probably far from the high-priced limos of the Argus set. That left us to mount the front steps of the farmhouse, brilliantly lit, not knowing what to expect.

To my dismay, we were the first to arrive. A flunkey in a red spencer jacket, crisp white shirt, and black tie, motioned us to the front door and with a smile directed us into the nearby living room, itself a symphony in luxurious furnishings, from the deep carpeting to the fortune in massive framed paintings on the walls. Here, I was astounded to see, was a reception line of the Argus Corporation's entire board and their ladies, including Mr. and Mrs. Cran. The men were in tuxedoes and the ladies in their most spectacular finery, probably to the last seam Paris-bought.

All eyes were turned to these first humble arrivals. Here we were, Madeline and me, facing a battery of Canada's most celebrated business tycoons and their wives, entirely unprepared for the grand event that was to follow. We'd never been used to hobnobbing with people bordering on royalty, as these seemed to be. But we somehow navigated the line, smiling and bowing as we shook hands with the mighty. I introduced us as we went. Cran smiled broadly, no doubt greatly amused at the overwhelming effect of the line on this simple minion and his spouse. Mercifully and swiftly, it was over. We passed on through French doors and out to a wide stone-floored patio loaded with pots of flowers. Others came in behind us.

From then on, the evening was fabulous. This was the era when corporations and politicians laid on festivities with a trowel. It was considered good public (and advisable employee) relations to toss a fete every now and then. I learned later that Argus was putting on a week-long series of receptions for personnel of the several companies they owned or controlled. They'd already done Dominion Stores and Domtar from Montreal. Tonight was Standard Radio's turn.

This was truly a spectacular event. Around at the back of the farmhouse was an enormous tent that would have served a small circus as its Big Top. The grounds around a swimming pool like the one in front of the Taj Mahal were impeccably groomed. Rose trees in little tubs stood around here and there. I noticed a cigarette butt nestling down beside one on the ground but almost instantly a gardener was there with a spike on a long stick and speared it.

The women guests, sparkling in their finery, found a reason to visit the house's facilities and got to see almost all its interior furnishings right up into the third floor bedrooms of the MacDougald girls. Only the best creamy rugs, pink-upholstered Hepplewhite furniture and shining candelabras graced these rooms. As the women gushed later, they'd never seen such luxurious insides, and toured as much of it as they could.

Beyond the Taj Mahal pool was a long white garage, open all along its front. Inside was a squadron of restored vintage cars, dating from a 1930s Auburn Speedster, a 1926 Alfa-Romeo, a Duesenberg or two, and on through 1930s classics that would be million-dollar prizes to car buffs today. A groundskeeper told us about the crushed-stone road that Mr. MacDougald had had built to wind through his acreage over hill and dale on which he personally drove his antique treasures every morning after breakfast. Two white-coveralled mechanics worked full-time keeping the cars in top performance shape. All we could do was goggle at the array in the garage.

There was a great roar from above, and the CFRB helicopter with Eddie Luther aboard swooped down and landed on green turf near the pool. Eddie and his wife hopped out to great cheers from the guests. The chopper took off again. Two sleek Windfields Farm horses, each bearing a young lady — presumably MacDougalds — galloped through the crowd and vanished over a hill. Bill Gilmour, one of my buddy newsmen, who was from the West, snorted at their riding style — *posting*, he muttered. Out West they rode in a more honest, Western way.

But the best was yet to come. A tuxedoed waiter moved among the crowd, muttering that we were invited to the great tent. We strolled into the Big Top. There in front of us, instead of clowns and circus creatures, was an acre or so of round tables — each topped with a pink tablecloth and a vase of red roses. There must have been a hundred tables. The dinner of many kinds of meats and exquisitely cooked veggies was dispatched with dessert of Cherries Jubilee, glazed and topped with whipped cream. Choice of red or white imported wine was ours.

W. C. T. Cran, with E. P. Taylor and Bud MacDougald at his side, explained to the gathering that CFRB and CJAD had great days ahead of them, and that there would be electrifying news coming soon. That produced puzzles for the mind. Hours later, heads spinning from the superlative evening, we shook hands with the MacDougalds and, along with most of the CFRB staff, took our places on the front porch to await delivery of our cars. One of the first to drive up was the seven-year-old Murray 1954 Ford, a symphony in rust, for all to see. Madeline and I dashed to it as quickly as we could, and took off. We'd had a staggering display of what radio could do. A week later, we bought a new 1962 car.

parade of the fading stars

Still, as there is bound to be, there were changes and a departure or two. The station's longtime singing star and director of music, Wishart Campbell, came into the newsroom one evening when I was the sole occupant to say a friendly goodbye, although as a newcomer I barely knew him. He'd been with CFRB since the early 40s, when live entertainment was an RB specialty, and was a reputable radio personality himself. He'd been one of the presenters at the BBG hearing, telling how CFRB-TV would air the finest musical programs. For him, it was never to be.

Now, with all hope for a TV career gone, he was "retiring." Where to? Well, he was going to marry a Mrs. Murray, a widowed Scottish lady on the island of Islay, just off the west coast of Scotland, where, it so happened, his bride-to-be owned *two* scotch whisky distilleries. He would help her manage them. No doubt he was secretly the envy of all CFRB staffers for this. I could relate to it in one way, as my mother had grown up on Islay, but I'd never been there. I wished him luck and we shook hands.

The newsroom seemed to be a temporary sanctuary for every variety of person who came to CFRB for whatever reason, either to be interviewed on-air or await the hour of their own program's start. I was working an afternoon and evening shift, and usually turned out to be resident host. The prestigious John Collingwood Reade — whose English tones had held a daily vigil on the progress of the war during the 40s, predicting this and that probable turn in strategies and events — came to sit and talk and smoke as I assembled news for the on-air man and "The World Tonight" newscast.

Walter Kanitz, dour proprietor of the evening's foreign music program, would saunter in next. I would learn that he'd fought in the Spanish Civil War. Ruby Ramsay Rouse, the organist, florid and expansive, might drop in and muse about many things, especially how audiences had changed from the lovers of organ music in the 30s to lovers of guitars and electric pianos.

Jolly Ray Sonin, popular purveyor of British news and music, beyond the pale today, famous in the U.K. for his variety productions on the BBC before emigrating to Canada, would come in, hunt for British news items on the teletype, and maintain lively talk. Now and then, Johnny Wayne, of Wayne & Shuster fame, would appear and chat with some of the old hands he'd known during the defunct days when RB did live studio-theatre shows.

One evening when Canada's centennial celebrations were coming up, a fellow came in while I was typing a news story, and asked if he could use the studio next to the newsroom, "to blow some of the nants out of this." "This" was an over-length trumpet with a tangle of embossed decorations sprawled along it, the like of which I'd never seen on a musical instrument. I said okay, and went back to typing. The fellow was Bobby Gimby, a well-known local bandleader who'd been commissioned to write a Canadian centennial theme song and play it at upcoming celebrations. The song was "Can-a-dah!" which became a children's chorus sensation and Canada's Expo 67 theme song shortly thereafter. I often wondered what became of that song after Canada's hundredth birthday was forgotten.

In the course of time some famous figures of the day came by. Chris Chataway, the world famous Australian track star, dropped in. Alan Small, our on-air classical music program man, brought in with him Lenny Bruce, the doleful U.S. comedian who became a cult hero to some. We chatted about nothing in particular. I think the poor fellow was under the influence of some kind of medication that made him rather vague and disoriented. Not being a follower, I had nothing to ask him about his latest or planned exploits. We drifted apart.

Madeline and I had the unique chance of meeting Bing Crosby, his wife Kathy, and their three children of Bing's second family, at the time in their teens or sub-teens. CFRB was one of the sponsors of a Red Cross benefit that included a videotaping of a Crosby special TV show at a Toronto theatre before an invited audience. Some of us were asked to be ushers for the guest audience.

I was really a fan of the Crosby of the early 1930s and wasn't especially fond of his later career, although I'd always found his films entertaining. The taping for the one-hour special took all evening, with frequent breaks. Afterward, at a green room reception, Crosby praised Toronto and sang some insider songs. He also nearly fell off a coffee table he was standing on. That would have been a major show-biz news event, but it didn't happen. He regained his balance. Crosby was surrounded by autograph-seekers, so I didn't get a chance to talk to him about his early days in radio and show business. We just got to shake hands.

After we were located in our sleek new quarters in the new building at St. Clair Avenue and Yonge Street, more celebrities visited the newsroom as I struggled with stories for the next newscast. Charlton Heston, Hollywood's Moses, dropped in to be interviewed by Betty

Kennedy. We passed in a hallway and nodded. Ginger Rogers was another, but I only got a glimpse of her.

Probably the blockbuster of all, one day Winks Cran came into the newsroom with a short, tubby, balding man in tow. This turned out to be E. P. Taylor himself, whom I'd already met at the MacDougald gala. Cran asked me to explain for Mr. Taylor the equipment we used for recording phoned-in news reports and newscasts from the American network — by this time NBC Radio, not CBS anymore. I began explaining to Mr. Taylor how we used the two big floor-based Ampex tape recorders, and how our equipment was all tied in with the on-air control room.

At that point, Cran excused himself and left to take a phone call. I was thus left alone with Canada's most famous — and probably wealthiest — businessman, the corporate boss-man of CFRB, whose name and picture often resounded through the newspaper and TV world, praised by many and vilified by others. I found him a quiet, ordinary guy. We chatted about the equipment as if he were interested in buying it, even though he already owned it. Cran returned. Mr. Taylor offered his hand, which I shook, and he thanked me for the rundown. They left. Slightly staggered, I got back to editing tape.

As a final grace note to the unspoken but suspected cause of CFRB's denial of the licence — i.e., the spectre of TV advertising of Argus Corporation's beer — within a year, the Ontario government, as mentioned earlier, lifted the ban on radio and TV beer ads. They even began to appear on the new CFTO-TV.

Whether this next meeting topped them all is a matter of opinion. As RB's first Ontario legislature correspondent I, with some other reporters, was invited out to the side entrance of Queen's Park to the Lieutenant-Governor's suite. The visitor: Prince Philip, the Duke of Edinburgh, all by himself. He arrived on the landing of the porte-cochere's stone steps smiling broadly, shook hands with us all around, then entered for late evening drinks and dinner with His Honour. We weren't invited in.

blast from the iconoclast

The news was beginning to develop along international lines in a way it hadn't since the Korean War. In 1959, U.S. President Eisenhower remarked that for the first time in decades there was no war going on anywhere in the world. That changed. Perhaps it had nothing to do with the arrival of President John F. Kennedy in the White House, but after the 1960 presidential election, Ike's remark was soon obsolete. In a tiny country in Southeast Asia, previously of little interest to the West — until the French colonialists lost it to the Viet Minh — trouble appeared.

Where there had only been U.S. "advisors" to the government working in South Viet Nam, there were suddenly fifty thousand U.S. troops being landed there. More were to come. An African Congo political rebel, Patrice Lumumba, a favourite of the Soviets, was assassinated, setting off a hot African war. In the West, the Cold War was getting colder. President Kennedy and Chairman Khrushchev weren't getting on. It all got worse. For the news media, the new crises created an unexpected boon. For CFRB, there was more than a boon: there was Gordon Sinclair.

Not known as an astute observer of international affairs, Sinc usually took the most unpopular stand on issues he often created himself, such as his daily radio rants against the fluoridation of municipal water supplies. He called the stuff "rat poison." Gordon didn't have any real scientific basis for his stand, and there have been no reports since of anyone dying from fluoridation, but that was Sinc's style. You didn't get attention by just agreeing with everything that came along.

People listened when Gord Sinclair denounced some current issue in his own winning way. His weather reports were brief and to the point, as "It'll be like a baby, folks: wet and windy!" His stock market report was just as brief: "Buy Tampax — it's going up!" His whole delivery, on-air and off, was like the exuberant style of jazz pianist Fats Waller: "Hey! Don't let it bother you if skies are grey! Laugh and be merry! Hey hey hey!"

The private Gordon Sinclair hadn't had much to laugh and be merry about during his flamboyant career. His only daughter died young in a Toronto hospital, and he never forgave the doctors who attended her. Doctors, in the Sinclair lexicon, were frauds and charlatans and should be shot. The loss of his daughter made him an embittered man, and it all came out in his view of the daily news of the world. It became known that he and his wife of many decades didn't get along. In fact, he had their Islington house divided in two, one part for him, one for her. Yet, at CFRB social gatherings, Mrs. Sinclair was present on his arm.

Financially, all was not well — even though Sinc had invested fortuitously in booming Laura Secord Chocolates stock, Tampax, and in Standard Radio. In the late 1960s, Sinc's investments took a serious dive when the market plunged. But he went on as usual, blasting his targets over the air and playing cribbage for money with old cronies in a back room of Stollery's Men's Clothing Shop on the corner of Bloor and Yonge Streets, where he would initiate the game by plunking a forty-ouncer of scotch on the table.

Sinc, contrary to his public image, was a reserved individual in his everyday life around the studio. Never appearing to be perturbed about much, with a peaceful smile and an open manner, he would come into the newsroom an hour or so before his next newscast and begin putting it together. I inadvertently got an insight into his script methods. This was through a side-job of editing a one-minute show called "Sinclair Says," snips from Gordon's eleven-fifty a.m. ten-minute daily newscast.

The idea was to lift these snips from the tape of his newscast and re-record them on cartridges (tape cassettes) for use on the air throughout the afternoon and evening. Early in the afternoon, I would run the Sinclair tape on a newsroom recorder and listen for items that would lend themselves easily to "Sinclair Says." Before starting the hunt, though, I would read his newscast script to help spot possible candidates. That was where I got my insight.

Sinclair's news script was a hodgepodge of items clipped from newspapers and magazines and pasted down in a random order on paper, with some Sinclair comments typewritten in between. On the surface, that seemed like plagiarism. However, on playing the tape, I found his actual on-air delivery had very little resemblance to the script. He simply ad-libbed around all his clipped items in his own terminology.

The end result was totally original Gordon Sinclair, no more plagiarism than the reading of any printed copy off the newswire or out of a newspaper. Besides providing me with frequent laughs, not to say confusion, in trying to locate the matching snip on the tape, it was a revelation of a renegade personality at work.

A cold war of sorts came to the CFRB newsroom one late afternoon in 1962. It was precipitated by an actual East–West Cold War standoff: the nervy attempt by Soviet Chairman Khrushchev to supply Fidel Castro in Cuba with an arsenal of ballistic missiles, known forever after as the Cuban Missile Crisis. In spite of U.S. President Kennedy's direct warning to the chairman that if he persisted in his attempt he could be precipitating World War III, the crisis went on for days. U.S. and Canadian radio and television networks were broadcasting hardly anything but the latest minute-by-minute play-by-play of the crisis, and newspapers were following up. Of course, radio newscasts were filled with it.

There was one disbeliever in the news spectrum: Gordon A. Sinclair, CFRB iconoclast and chief pooh-pooher of many things in life. Predictably, the Cuban Missile Crisis became one of his pet targets. Gordon didn't believe for a minute that Chairman Khrushchev would actually carry out his mission. It was all a bluff, Gordon said. In his two newscasts a day he railed at everything and everyone connected with the "crisis," brushed it all off, and went on with "The News."

There was a crisis of another kind brewing in the CFRB news operation, starting well before the Cuban issue, and Sinc was involved in that, too. His power at CFRB was considerable. He was regarded by management as one reason that the radio station was as popular as it was. They didn't want to offend him. Some of the listeners loved him; others hated and despised him. But they all listened to him.

He held such an esteemed position at CFRB that he could warn the management with impunity that if he were ever cut off the air for some news bulletin there would be trouble they couldn't imagine. He would denounce them all over the air if necessary. Gordon also happened to own a large number of shares in Standard Radio, the owner of CFRB.

Not long before the Cuban crisis, our news director, Bill Hutton, had become intrigued by a new gadget being pushed for its affiliated stations by CBS News. This was a black box with a little curved window in it where numbers clicked by one at a time, second by second, when there was what CBS News considered an urgent news bulletin. It sat on a newsroom shelf and gave instant access to the CBS line and a CBS newsman in New York who would, with a bare ten seconds warning, rap out the bulletin. The warning came in a series of ten clicks.

Hutton convinced Winks Cran that this black box should be wired into the control room with a button in the newsroom to push as soon as it started to click. There was to be no question that when the CBS black box clicked, the newsroom man was to push the button before ten seconds were up, bypassing the control room, cutting the current program off the air and instantly launching the CBS news bulletin — in all its burning-issue glory — onto CFRB airwaves.

Bill, in all sincerity, believed that the black box would give CFRB News a giant jump on all other local stations with its instant bulletin capability. He directed a memo to all newsmen and explained how, if any were in the newsroom when the black box started to click they must rush up to it and push the button before ten seconds passed, regardless of what was on the air at that moment. There wouldn't even be time to warn the on-air control room operator that a bulletin was coming and that the current program would be cut off, even if an announcer were in mid-sentence. Bill was extremely firm about this. No one was to be derelict in pushing the button instantly, and damn whatever was on the air.

In a quick meeting in Hutton's office, I suggested that not all CBS bulletins had vital interest for CFRB listeners, even though they might be vital to CBS listeners in the U.S. Why not, I ventured, *record* the bulletin on tape when the black box coughed one up, and then, if the newsroom's news sense deemed it worthy of breaking into a show, do so with the tape. It could be rewound in seconds and a warning could go to the on-air person that a bulletin was coming. CFRB would still get the jump on its rivals. That was when I got my journalistic fingers burned.

With great vigour, Bill vetoed the idea out of hand. The system was designed to pre-empt any programming no matter what, and it had to be used that way. I asked what might happen if a bulletin came in when someone like, say, Gord Sinclair or Jack Dennett or Betty Kennedy was on the air. Hutton vehemently said it didn't matter a ******* *who* was on the air, *push that button!* The order was in concrete.

Naturally, when the Cuban Missile Crisis was at its height, Sinclair was on the air with his five-fifty p.m. newscast, holding forth on the phoniness of the U.S. media's frenetic play-by-play coverage. I was putting together some items for Jack Dennett's possible use on his six-thirty p.m. newscast, when what I'd been anxious about happening happened. Sinclair was halfway through his newscast. The black box started to click. When ten seconds later the numbers reached ten — the magic signal — and my finger hesitated on the button. I cracked on the tape recorder.

Not known to me, a duplicate black box was in Mr. Cran's office. When it clicked in the newsroom, it also clicked in the boss's office. Also unknown to me, both Mr. Cran and Bill Hutton happened to be in that office together when this particular clicking began. Apparently, Bill waited proudly and confidently as he and Cran listened to Sinc's newscast on the office radio, expecting to hear the CBS newsman cut in with another earth-shaking bulletin.

When it didn't happen, Bill, no doubt put off by some derogatory remark from Cran, charged furiously down to the newsroom, bouncing in on me with his face crimson and his voice savage, demanding, "Who the hell didn't push that button?" Of course, I said, "I didn't. I taped it."

The castigation that followed will remain unreported here. The upshot was that I, recognizing the enormity of my delinquency, nodded many times and pledged to push the button the next time the black box clicked. I felt I'd done my part, but it was lacking. Hutton repeated his order, still angry, then swung around and left.

Next evening at the same time when Sinclair was in mid-denunciation of the "crisis," the black box began clicking again. Only following orders, at number ten I pushed the button, and mentally ducked. I knew what was coming.

The CBS announcer started in with his high-urgency style to tell us that a Soviet ship was now halfway to Cuba, something he'd told us only a half-hour before. Was this a development important enough to put all programming off the air? I didn't think so. But it was too late. I'd punched the button. The CBS newsman went on the CFRB air, and Gordon Sinclair was chopped off in mid-word. I waited for the storm. It didn't take long in coming.

Sinclair hurtled into the newsroom, still clutching his script, and demanded to know who'd cut him off the air for one of those "!'@%*+@'d American news bulletins! Laying my head on the block, I said, "I did." I felt like young George Washington caught with his little axe in hand right at the cherry tree. He stopped short. Then he said, "Okay, I don't blame *you*, Gil. You had to do as you were told. But I'll get the bastard who ordered it!"

Then he stormed away back to the studio to wait for the end of the bulletin. When it did end, the control room operator, also ducking the flak, snapped the switch putting Sinclair back on the air. He caught him right in the middle of raving over the outrage, yelling something to the effect that "those god-damned bureaucrats will pay for this! What the hell was in that bulletin important enough to cut me off the air?" He didn't know he was actually back on the air, ravings and curses and all. When he did, he still went on in that vein, berating not only the Cuban Missile Crisis as a phony publicity game between Kennedy and Khrushchev, but also blistering the CFRB management for cutting him off the air. The "Crisis" was no crisis, Sinc swore. Khrushchev would back down. He was right. Ultimately, Khrushchev did back down.

That didn't end the *CFRB* crisis. Actually, it worked out well for Sinclair. The whole incident, Sinclair's on-air swearing and all, was reported all across North America on radio, TV, and newspapers. He became an instant continental media celebrity for not only denouncing the whole Cuban affair, but also for being right about its outcome, at that. Khrushchev called back his missiles at the last minute. Kennedy became a U.S. hero, seen as steadfast in staring down the Soviet headman. But Sinclair was heard everywhere, U.S. and all, startling everybody with his on-air language. The perpetrator of his outburst, Gil Murray, didn't get a mention. That was entirely okay with me.

Still, the RB crisis went on. I wasn't invited to sit in on the meeting that took place the next day in Cran's office. All the top hands were there, including Sinclair. Eventually I heard how he showed no mercy as he ripped into Bill Hutton and his black box and his order to use it no matter what. It seems Gordon won his point.

Nobody was surprised at that.

Next day a memo over Bill Hutton's signature was circulating among newsmen saying that henceforth all bulletins coming in on the CBS black box were to be tape recorded first, then instantly judged whether they were urgent enough to go on Canadian air within a few minutes. Not long after, the black box disappeared from the newsroom, never to click or be seen again. Nobody said anything more about it. The Sinclair–Cuban crisis was over.

the zanier, the better

Where do radio announcers come from? What prompts a young person to launch on to the airwaves and attempt to survive mostly on his/her charming self? The answers to both questions, and probably to others unasked, is that a high order of self-confidence — even of ego — is the driving inspiration. Oh yes, there is also the lure of Big Money. Sadly, not all of the thousands of announcers across the country receive the rewards the public assumes they get. Of course, the *stars* are extremely well-paid. That's another story.

Without the ego factor, it is doubtful that many DJs, talk show hosts, and opinionists would ever get on the air. In many stations, by management's concern for the station's integrity and for boosting its audience popularity and the effectiveness of its advertising, and by a strong degree of government regulation, radio must be an entertaining twenty-four-hour talkathon, with music pumped into the air between bursts of frenetic discourse. Dead air is taboo. Long (more than about a minute) silences might make listeners think the station has died, and switch to a competitor.

The advertising manager might have a fit, as seconds and minutes are valued in radio by extremely expensive measures. Dead air is dead money. The program director might jump to the conclusion that a lazy DJ is before the mike, and is resting. Others might assume the DJ has run out of things to say. He/she'd better have a good story ready for the boss if he/she is to achieve forgiveness. So the successful DJ, talk-show host, etc., must have what once was called the gift of the gab, be never at a loss for words, be ever lively, and always seem on top of the situation. Cool, cool, cool is the rule. A tranquilizer may be a vital career aid.

There is probably no activity beyond skydiving or lion-hunting that calls for as much presence of mind and quick reaction time as on-air performance. Only the television newsperson in fast-changing live news situations must have steelier nerves and a cooler head, because he/she must not only *sound* cool but *look* cool and contented as well. The radio newsperson has the advantage of glaring desperately into the control room for answers to a glitch while keeping his/her voice well-tempered on-air. That's where radio is even more of an illusory medium than TV. What the listeners can't see is what they don't get, and no harm is done.

The exuberant, polished Voice of radio was not always so. The very name "announcer" is a clue to what the Voice was like when radio was young. The word suggests a formal, highly proper person announcing facts with cold certainty and no nonsense. And, indeed, that was the accepted technique. After radio came to be seen as more than a voiced telegraph medium, there was a need to have it taken seriously by audiences. Announcers, even some technicians, at first wouldn't be seen in anything but a business suit, unless it was a tuxedo, as some did wear especially at public events like live dance hall broadcasts. On the air, chatty talk was not expected, except from the comedians.

When Lorne Greene read out the CBC ten p.m. news during World War II, you were to realize that this was serious stuff. When the classical music program followed, the mood was extended to impress the listener with the gravity of Mozart's "Jupiter Symphony" or Beethoven's "Fifth." Commentators spoke with sombre conviction to command rapt attention. The outbreak of glee when "The Happy Gang," or Jack Benny came on the air was all the more contrasted with what had gone before and would come back after the show was over.

Announcers of popular music programs also cracked the mould. When the music of the Big Bands was the program, especially if it were a live broadcast from a dance hall with the announcer standing up at the microphone on stage in front of the massed dancers, it was more appropriate that he hit the airwaves with a light party sound. Jovial, he brought the unseen essence of radio to the scene with one hand cupped over an ear, the other gripping a sheaf of papers from which he read out the name of the next piece the band would play. The hand on the ear helped give the dancers the illusion that they were part of a Hollywood scene, but it also let the announcer hear his own voice over the din so he could fine tune it to his normal quality tones. Otherwise he'd have to shout, losing all accepted radio decorum.

Something happened to all this in or around 1950. Although in the 1930s and 40s there had been radio mavericks such as Walter Winchell in the U.S., and Gordon Sinclair and Jim Hunter at CFRB in Canada, most broadcasters stuck to the more inhibited style that was general throughout most of North America. It may have been because of the new raucous role of rock 'n' roll that, in the 50s, the DJ appeared and inspired his fans to seek the frenzied heights of the new music's throb and shriek. Along with them, he threw off the reserve of older radio. From there on, wilder talk and tones gradually engulfed the Voice of radio, even invading the newscast with an earthy sound, setting the "on-the-scene" image.

Of course, a spur to joviality always existed in the bottle. As with many movie stars, over-indulgence in alcohol often brought down radio people, and not just among the big names. A prominent city hall reporter kept a bottle always at the ready in his station's city hall office filing cabinet. Reg Wilson, an old-time control operator at CFRB, used to tell of one newscaster who might or might not show up for a scheduled newscast — and if he did, there was no guarantee he'd complete it.

More than once, Reg claimed, while he was busy sorting records and tapes in the control room, there would come a dead silence on the monitor. When he looked into the studio, Announcer X was nowhere to be seen. He'd slipped under the table, out cold. Reg would rush a record on to a turntable to fill the gap, but as a non-announcer he couldn't take over the newscast on-air. He could only play more records until the DJ finally showed up. Operators formed the fail-safe habit of keeping at least one record on a turntable ready to go in emergencies.

In my days at CKTB as an announce-op, I had more than one escapade with on-air people's over-indulgence. Once, a guest sportscaster from another station, subbing for Rex Steimers who was away, was so blotto when he arrived that I was afraid he wouldn't make it as far as the little studio. He did, and I introduced him, on-air from the control room, perhaps too patronizingly, as our special guest.

For two minutes he mouthed a ragged delivery of the sports news from wire copy I'd hurriedly scrabbled together for him and shoved into his hand. Then he began to wobble, finally flopping across the table unconscious. Startled, I put on a recorded commercial spot and rushed into the little studio, where I grabbed the sports copy, rushed back into the control room, and with fake casualness read off the rest of the sportscast on-air, noting that Sportscaster X had had to leave to cover a hockey game. Nevertheless, X later went on to a successful broadcasting career.

There were former stars, losing their on-air appeal but promoted to prestigious symbolic off-air management jobs who couldn't take the new anonymity of not being heard any more, except perhaps in the boardroom, and that with indulgence by the others. That usually meant extra time on their hands, and therefore opportunities for late afternoon refreshments they hadn't had before. Altogether seeing themselves losing their magic, they failed to adjust and went downhill from there. The casualty rate among deposed celebrities was high. Sustained performance and popularity were crucial for remaining on top of the heap in radio.

charting new dead reckoning

In late 1960, with quiet having come to the Great Toronto TV Licence battleground, where next for Standard Radio? Would Winks Cran stay on? Had he been hired by Argus especially to get the licence? Apparently not. Cran still turned up often in the newsroom to talk with Hutton, and after a few weeks passed, he looked pretty much like his old self, unshaken, with no intensity of the campaign showing on him. Some of us wondered what would be done with the $2,134,129 plus $1 million bank credit earmarked for CFRB-TV. Would Argus just co-opt it and buy another business? We stayed tuned.

I'd noticed a story on the news wire saying that in Etobicoke, one of the Metro Toronto boroughs, a U.S. company was running a marketing test of a new and interesting idea: linking up several hundred homes to send them TV programs by way of telephone cable instead of broadcasting from a transmitter. It would be a closed-circuit service, like the one Gord Sinclair starred on at the CNE way back in 1939. Here was a new idea for television: Cable TV.

If Standard were to set up a system like that, a TV licence wouldn't be needed. I mentioned this to Bill Hutton, suggesting he take up the idea with Cran. If our leader went for it the idea would be a feather in Bill's cap. It could be just the place to put the multi-million dollar nest egg. Bill just smiled and said: "I'm sure if it's any good, he'll have thought of it." End of discussion. Apparently he didn't. No Standard Cable TV service ever appeared.

Not too many years went by before a cable company or two sprang up — and not just in Toronto — who went all-out in wiring up thousands of homes in just that way. The yagi-loaded TV towers beside most homes gradually vanished as people went for cable TV. Ironically, one of the biggest cable companies was to be Rogers Communications, run by Ted Rogers, Jr., son of CFRB's founding genius.

But that wasn't the end of the affair for Standard Radio. RB's program style was still going over in grand style with that solid majority of Metro Toronto listeners. The personalities were what did it. Wally Crouter was still the most popular morning man; Gordon Sinclair and Jack Dennett still held the biggest newscast audience with four hundred thousand *households* each; Betty Kennedy and her afternoon interview program was going out to about the same. Bill Deegan's "Rolling Home Show" still commanded the biggest late afternoon rush-hour audience; the evening music programs, such as Ray Sonin's and Walter Kanitz's drew good, if smaller, audiences too.

In the crucial Neilson ratings, CFRB's curve soared high above all others, with CHUM charting in well below it, and other radio stations, including the CBC, hardly more than a weak line along the chart bottom.

The expansion of the news department, started in the days of great expectations for the TV licence, stalled but didn't reverse. Bill Hutton, who seemed to be still in the best books with Winks Cran, continued as news director. One of his ideas was to take advantage of the CBS — later NBC — news connections, for which Standard Radio held the Canadian franchise.

Bill's late position with Canadian Press inspired a new CP arm, Broadcast News (BN), a service designed to supply radio stations across the country with the latest news voice clips, not only from CBS — or later, NBC — but also with reports phoned in to CFRB's newsroom by newsmen in those same Canadian stations, to be re-fed twenty minutes before each hour across a telephone network to the member BN stations. This gave them time to pick clips from fresh news feeds for their next hourly newscast.

The idea caught on quickly with the stations, almost all of whom were already Canadian Press members, and dependent upon the CP teletype for their constant flow of national and international news. Now, with actual voice reports coming in from NBC and stations across Canada where news was happening, such a service couldn't be resisted. Each reporter, except from NBC, got paid for each story phoned in to our tape-central in the CFRB newsroom. That assured a good supply to be re-fed in batches by phone lines to network members.

In effect, a CP radio news voice network, run by RB and years later turned over to CP's Broadcast News service, had been created, thanks to Bill Hutton. It proved lucrative to Standard Radio under the name of Standard Broadcast News (SBN). One contributing newsman, at CHNS Halifax, was the rotund Mike Duffy, of later CTV Ottawa bureau fame, getting his national reporting career under way. Other news reporters in stations coast-to-coast got national exposure as well.

Mulling over how we could expand our news coverage as originally envisaged before the TV licence fiasco, we could see a need for on-the-spot news reporting from the same basic sources that Toronto newspapers had used for a hundred years — Toronto city council and now also Metro Toronto council; the courts, from the Ontario Supreme Court down to police court; the police beat, handled by Jocko Thomas; the Toronto Transit Commission; and, what was to be my own pet news beat, Queen's Park, home of the Ontario legislature. If CFRB were to live up fully to its slogan as "Ontario's Authoritative News Voice" so often advertised, we felt, then the delivery of news from actual scenes of daily happenings in and outside Metro Toronto should be its aim.

In the traditional vein of the newspaper reporter, I believed there should be several reports each day from Queen's Park while the legislature was in session. So I had my chance to boost this source of important news that usually had only routine coverage in newspapers at that time and even less on radio and the new-hatched television stations. This, and coverage of other legislatures, was grist for the SBN network mill, too.

Early in my ten-year career at the *Toronto Star* I'd briefly covered all those news sources at one time or another. Radio, despite its ability to report news the instant it happened, had usually been late in catching up to the papers. And mostly it just gave out the skeleton facts, and only some of those, at that. Here was a chance to make the most of radio's potential for instant news reporting. We began to explore the possibilities.

Hutton and I dropped in at the Queen's Park press gallery. The news potential was obviously great, and largely untapped by radio. But we had no reporting staff. Although I'd been hired as CFRB assistant news director and editor, I suggested doing the reporting myself. Hutton gave the okay. It wouldn't cost much. RB joined the press gallery, was assigned desk space, and I began news coverage. Instantly, the regular hourly RB newscasts became peppered with brief but complete news items of importance to the Ontario listener. The tapes were broadcast again and again the rest of the day, and were transcribed to paper for use by Jack Dennett and Gord Sinclair and on the hourly newscasts.

At first I just phoned in the reports, sometimes from a telephone just outside the premier's office door. They were tape recorded at the station and usually used on the next regular on-the-hour newscast. As the new on-site coverage met with more approval from the RB management, Bill Baker's technicians installed a microphone and amplifier hitched to a broadcast-quality phone line from our "office" on the old legislative building's third floor. The sign-off "Gil Murray at Queen's Park" put an exclusive stamp on both the CFRB and SBN reports. Reacting, reporters from other stations began to turn up at Queen's Park, too.

After making many beelines from the press gallery overlooking the legislative chamber, up two broad flights of marble stairs, along a vast corridor to the open area where my desk and equipment sat, I began to wonder if I'd bitten off too large a chew. At thirty-six, three hundred yard upstairs dashes were not my favourite workout. But as the essence of reporting news is speed, it did seem worth it.

One reward was the comment from a friend who still worked at the *Star* who mentioned there was now a radio in the *Star* newsroom, usually tuned to CFRB. When one of my one-minute reports, fresh from Queen's Park's house sitting, came over as the newspaper was about to go to press, Tommy Lytle, the *Star's* acerbic news editor, would yell at the city editor, "Hey, Gil Murray's scooped you again!" There was a lot of satisfaction in hearing that. Radio had scored big.

All Toronto newspapers — and the *Spectator* in Hamilton, the *London Free Press*, the *Windsor Star* and Canadian Press — had fulltime reporters based at Queen's Park. The *Star* and the *Toronto Telegram* had three each, the *Globe and Mail* one. CHCH-TV in Hamilton had frequent reports from freelancer Don O'Hearn, and CBC-TV had Art Robson. But the only other *radio* station having on-the-spot reporting, and that just occasionally, was CKEY Toronto, done by its dedicated news director, Godfrey Hudson, who dropped in now and then for brief news pickups. Later, Bill Rathbun covered for CKEY after Godfrey became seriously ill.

Even the new CFTO-TV had one of its reporters, Larry Kent, come around with a cameraman at times. Ironically, he often leaned on me — the one from the failed TV station — for tips. CBC Radio assigned Bill Murphy, and sometimes Ed Cosgrove. Previously rating only an inside-page column in the *Toronto Star*, "Up at Queen's Park," by old-time reporter Roy Greenaway, and something similar in the *Telegram*, the Ontario legislature was becoming a hot news source by 1966.

At the time, there were no "corridor scrums" with cabinet ministers or the premier as developed later both at Queen's Park and Ottawa. There were no press conferences either. On the advice of former *Globe and Mail* reporter Bill Kinmond, Premier John Robarts' communications director, Robarts grudgingly agreed to meet directly with the news media. Kinmond called a press conference to introduce the new

— in 1963 — premier (or prime minister, as the top man was known then). News reporters and TV cameramen bearing huge film (not yet videotape) cameras on their shoulders, and flashing hot halogen spotlights around, jammed a small room used by committees near the premier's office. The air inside was hot and smoky.

Robarts, looking grumpy about this news nonsense, could hardly push his way through the mob to get to a big polished table at the far end. When I placed my solitary microphone on the table in front of him while the news people crouched about in chairs or on the floor, Robarts made a sweeping gesture with a wave of a hand and boomed: "Get that damn thing off the table!"

I snatched the mike back instantly. Kinmond rushed over and spoke into his ear. The premier then nodded resignedly and I was allowed to put the mike back on the table. I was bold enough to assure him that only his announcement would be taped along with any answers to reporters' questions. With the passage of time, the same table was to bristle with radio and TV microphones as Robarts and his politicians caught on to the usefulness of the news media in getting their messages across to a wider audience. At last radio was recognized at Queen's Park as having a legitimate place in serious Ontario legislative news coverage.

# of shoes and ships and sealing wax

And the time had come to speak of many things. With CFRB in place at Queen's Park just in time, news that was to have direct effects on the pocketbooks of the listening audience began arriving in bursts. Leslie Frost had just retired as Ontario Premier and the new government under John Robarts was soon bringing in unprecedented legislation. There was gold-plated news to mine in the universal health plan called at first OMSIP then OHIP, the Ontario Health Insurance Plan; as well as permission for unionized teachers, police, and firemen to strike if they thought it necessary.

There was more money for roads and expansion of superhighways such as Highway 401, renamed by the nationalist-minded Robarts the Macdonald-Cartier Freeway; new laws on gambling; founding of the Department of University Affairs; and other landmark legislation. It was a kind of news everyone could grasp quickly, almost tailored for radio's short reports, with taped inserts of the newsmaking politicians' actual voices. It was the beginning of the era of a need for more rapidly transmitted news and information happening all across North America.

In 1963, one of the most sensational scandals involving the Ontario government popped up, resulting in Robarts appointing the Roach Commission on Organized Crime in Ontario. The weeks-long hearings on all gambling in Ontario resulted in the switching of one cabinet minister, Attorney-General Kelso Roberts, to the ministry of lands and forests — in effect, being sent to the wilderness — and naming of an obscure MPP, Arthur Wishart, to the justice post. It also revealed that gambling syndicates were operating betting clubs clandestinely under old bought-up provincial charters, like the one for "The Ontario Birdwatchers' Club" and "The Cairney Road Quilting League," with a politician or two looking the other way.

All these developments made news of a kind that radio stations had previously simply "scalped" skimpily from newspapers and wire services. I covered the weeks-long hearings, in which some of the shadiest characters I'd ever seen — even in the courts — behind black sunglasses and racetrack touts' loud jackets, testified on the underside world of off-track gambling. They looked scary enough for a Hollywood mafia movie.

Eventually, CFRB — and radio in general — received recognition as important news outlets with your humble author's unexpected election as Vice-President of the Ontario Press Gallery, an official provincial entity under the umbrella of the Speaker of the House. In 1965, after the assignment of the *Telegram*'s Fraser Kelly (sitting president) to the *Tely*'s Ottawa bureau, I became president. Thus, with the first radio man — a former newspaperman at that — to be elected president of the Ontario Press Gallery by the mostly newsman members, radio people, formerly thought rather frivolous, now gained status on the same plane as newspaper reporters.

This tended to break Canadian provincial patterns of exclusively electing newspaper reporters to head press galleries. The status of broadcasting legislative news was rising among the public and politicians, as more and more live coverage of news evolved in radio, and some time later in TV newscasts. The patterns set by radio for on-the-spot coverage largely were followed by television newsrooms, as TV began to emerge as a serious news broadcasting source, but with pictures.

The public got faster and more detailed original reports, while the politicians got to speak right to their voter targets. The newspapers were briefly set back, but it prompted them to beef up their coverage of provincial, not just federal, news.

Queen's Park news always dried up temporarily when the House finished its sittings — in those days longer and more frequent than in later years, with lengthy fall and winter sessions, even, at Robarts' insistence, sitting at the House at night rather than in hotel rooms — so in the gap I would switch to Toronto City Hall and Metro news coverage. The then-new Toronto City Hall, the great clamshell, was just going into use. I'd spent many hours covering city council and the courts in the old ornate city hall at Queen and Bay, a building I admired but that depressed me by its gloom.

Now this solitary radio reporter sat in on the Toronto council's inaugural meeting in its bright new home. From then on, sparked by revived civic enthusiasm, Toronto's skyscraper era of the coming 1970s and 80s rapidly approached. The upsurge was also inspired by the international attention brought by Finnish architect Viljo Revell's futuristic design that won the 1958 world contest for a new Toronto city hall. The results of the 1960s city revamping plans, for better or for worse, can be seen now in Toronto's downtown forest of steel and glass towers.

One of the first obvious things lacking in the new city hall was any accommodation whatsoever for news coverage by radio and TV reporters. The newspapers were already well fixed up with office space for their operations. In keeping with the sparse live news coverage given to city council news by radio and TV, benighted city hall bureaucrats hadn't bothered to make room for broadcasters. They apparently assumed broadcasters got all their news from newspapers. Some did.

CKEY's Bill Rathbun and myself requested an audience with Mayor Phil Givens and got him to agree that broadcasters, too, needed a city hall base equal to what the newspapers had. With his worship's go-ahead, we found space in the clamshell's sparkling basement that we could divide up between us.

City hall builders put in partitions; CFRB's, CKEY's, and CBC's technical people installed mikes and amplifiers as at Queen's Park; and Metro Toronto listeners were assured of regular radio reports on what their councils were up to. The politicians, able now to speak directly to their voters on tape, loved it. Soon other radio and TV stations acquired cubbyhole office space. A new era of broadcasting news of Canada's biggest city was launched.

There seemed to be something revolutionary for Toronto every day the city and Metro councils sat. As more radio reporters, and finally TV crews, appeared regularly on the scene, Toronto and Metro news began reaching audiences almost instantly, jarring newspapers into a different, more magazine-like approach to their Toronto coverage. Their trend began to turn to more expression of opinion through columnists rather than by straight reporting of municipal news happenings. In other Canadian cities the same trend to more active broadcast reporting was happening. The young man or woman with mike in hand was more and more becoming a routine sight all over the country, to the point of news "scrums" looking like mini-riots.

An upgrading in the public's (read politicians') view of radio news came with the Toronto election of 1964. Usually, the candidates in any election and their publicists worked hard to get as much story and picture space as possible in the newspapers and TV. Radio, regarded as just plain entertainment, had always been a second- or third-place runner in the news game. Times were changing.

Bill Hutton let me know late one afternoon that he wanted me to join up that very evening with Allan Lamport, running for City of Toronto mayor for the umpteenth time, and do the rounds with Lampy of candidates' meetings, where the voters were invited to hear from the many election hopefuls. After one week of that, I was to switch to the campaign of incumbent Mayor Phil Givens, and stick with him for a week.

I suspected that this arrangement had been made at the request of the politicians themselves. Knowing the publicly explosive Lampy well from his past city hall exploits — once mayor and a longtime controller years before, and later chairman of the Toronto Transit Commission — I expected some interesting fireworks.

Such public meetings were arranged by the League of Women Voters of Metropolitan Toronto in school auditoriums, church halls, even private residences. I arrived for Lampy's first public showing at a church hall crammed with voters. He was onstage, keyed-up to start his pitch as soon as he was introduced. Of average height but of portly girth, a reasonably good-looking man of sixty-four despite his excess weight and balding pate, he presented for the public group a familiar image. He seemed always on the verge of smiling, but maintained a reserve of seriousness in keeping with his political persona as "your dedicated public servant." He was a familiar sight, having been end-lessly lampooned in newspaper cartoons, for which he was a perfect tar-get, and a favourite victim of the news media.

At that first Lampy speech I took copious notes on his promises, expecting to gather from him a proliferation of lively ideas and tech-niques as time went on. I was to find, as we dashed from one school or church hall to another, in Lampy's rented, chauffeured limousine, that he would repeat the almost identical speech word for word at each place, even at a ladies' tea in a private apartment, using no notes and at times fumbling on his words winningly.

I had it all in my memory after the fourth or fifth engagement. It seemed redundant for me to go on until nearly midnight with the emi-nent candidate — I was the only newsman with him — but I had my orders, and we tramped into crowded halls on a pre-laid circuit until I lost track of where we were time after time. Once, around suppertime, we entered a banquet hall with tables filled by some kind of club's members. Lampy went directly to the podium, was introduced, made his speech, grabbed a couple of bites to eat at the head table, then bowed out. I was-n't even given a seat at a table, let alone anything to eat. That didn't seem to bother Lampy. We rushed off for the next engagement.

Lampy, never really loosening up with me or cracking his facade of dedication to the public good, did his thing time and again. He seemed to have put himself into a mode of sober purpose befitting an aspiring mayor, with no time for the surface joviality usually expected of politicians. He even projected an air of humility. After the evening's last speech around about midnight, Lampy got out of the limo at his own house and told the chauffeur to take me to CFRB. I taped a roundup report for overnight and next morning newscasts.

After four more nights of Lampy-talk, the same theme over and over, I was hard put to find something new to report on-air each night. I was convinced I knew everything that Allan Lamport was likely to say for the next two years, if he were actually elected mayor. He wasn't, but he managed to remain in the public eye as an ordinary alderman almost until the day he died at the age of ninety-eight.

Next, I dutifully joined up with Phil Givens. I knew him from our weekly interview meetings for CFRB's "Office Of The Mayor" program. Now we travelled in the official limousine of the Mayor of Toronto. Givens' personal political aides came with us, and sometimes there was also a reporter from a newspaper. Hurtling around the dark streets of Toronto for five consecutive nights from one hall to another at top speed, running red lights, in a long, black, darkened limousine filled with big, mostly silent men in fedoras and overcoats, bent on annihilating the opposition, was uncomfortably like being caught in the making of a Hollywood gangster movie. Once our chauffeur, over-enthusiastically shooting the limousine away from the curb in front of the Toronto Board of Education building, bounced us off a car just coming out of the building's driveway. It was a minor crash, but Givens told the chauffeur to give his licence number to the unfortunate citizen for him to make a claim, then get going. The collided citizen was probably stunned when he found out he'd been creamed by the Mayor of Toronto.

We did the rounds of the same halls I'd known with Lamport. Once, we arrived just as Lampy, still on the campaign trail, was winding up his speech there. The atmosphere as the two of them passed in the aisle was as frosty as the air in the Arctic ever gets. Not a word nor a smile was exchanged. That's politics. When my five-evening stint with Givens wound up, I was, as usual, let out at the CFRB Bloor Street studios. Glad to finish up this assignment, I jumped out of the mayoral limousine with no delay and darted into the RB building. The mayor left.

Next afternoon, working in the newsroom, I had a call from Jean Marno, our accomplished receptionist, who told me a man in a uniform wanted to see me. I went out to the reception desk. Phil Givens' chauffeur stood there, holding my hat and coat, which I'd inadvertently left in the mayoral limo when I'd leapt out the night before. It was some time before the staff stopped marvelling about how the Mayor of Toronto had sent his chauffeur and limousine to deliver Gil Murray's hat and coat! As for me, I firmly believed I'd earned every bit of the honour.

Political news, with all its importance, was not the whole story of the expansion in the developing coverage by radio news services. Radio reporters were beginning to pop up wherever there was something or somebody making news. As television, bringing moving pictures into homes, gained more power, radio everywhere was obliged to improve its news services in its own defence. The old "rip-and-read" practice of simply parroting what came over the newsroom teletype no longer would do. Radio news needed the spice of live reporting from actual newsmaking locations. A new field opened for journalism graduates.

The time gap between a news break and a major news figure's reaction was being squeezed down. Now, instead of days passing between the two, instant comments were forcing a more aggressive news service. Immediacy of what was happening, whether beneficial to anybody or not, began to speed up the pace of life, even though the general public might not have been aware of it.

Radio was there.

Gallery of Radio Giants

A former president of the University of Alberta, Dr. Andrew Stewart was appointed in 1958 by Prime Minister Diefenbaker as chairman of the first independent regulatory body on Canadian broadcasting, the Board of Broadcast Governors. Creation of the BBG ended the objection of private broadcasters to regulation by the CBC, which competed with them for advertising, personnel, and broadcast frequencies. Dr. Stewart repeatedly declared that he knew nothing about broadcasting. But it was believed that his service on various royal commissions under different governments would assure his impartiality. One of the BBG's more controversial decisions was the award of the last available TV channel in the lucrative Toronto market to the *Telegram* newspaper in 1960, after keen competition among nine media outlets, mostly broadcasters.

Apppointed first chairman of the Canadian Radio-television and Telecommunications Commission (CRTC), replacement for the Board of Broadcast Governors in 1968, Pierre Juneau was responsible for the ruling that all CRTC-licensed outlets be at least eighty percent Canadian-owned. Under him, the new governing body created a Canadian content policy for music programmed on radio stations, that thirty percent must be Canadian. The Juno Awards for Canadian recordings were named after him. He was appointed Minister of Communications by Prime Minister Trudeau in 1975, but failed to win a parliamentary seat in a by-election. In 1980, he was named deputy minister in the federal Department of Communications and Undersecretary of State.

The first Newfoundlander to be a member of the Ottawa Press Gallery, Don Jamieson rose through a radio station partnership with Geoff W. Stirling to President of the Canadian Association of Broadcasters (CAB) and, by 1966, to election to Parliament. As a young Ottawa reporter in 1945 he covered the proceedings that resulted in the entry of Newfoundland into Canada in 1947. From 1968 on, he held several federal cabinet posts and finally was Secretary of State for External Affairs. Later he was High Commissioner for Canada to London. He returned to Newfoundland and acquired full ownership of the radio stations he and Geoff Stirling had jointly owned.

A son of Sir Clifford Sifton (a minister in the government of Sir Wilfrid Laurier), Clifford, Jr., was a strong figure in Western Canadian broadcasting. In 1928 his family had acquired the *Regina Leader* newspaper (later the *Leader-Post*), and by 1953 he was its sole owner, and the owner as well of the *Saskatoon Star-Phoenix* and radio stations CKCK, Regina, and CKRC, Winnipeg. He was also a shareholder in CKOC, Hamilton, Ontario. He formed Armadale Communications, taking full control of CKOC, but selling interests in CHCH-TV, Hamilton. Armadale began Regina's first TV station, CKCK-TV, in 1954, and later obtained licences for FM stations in Regina, Winnipeg, and Hamilton. For many years Sifton was on the Board of Trustees of the Canadian Association of Broadcasters.

Roy Thomson was unique in Canadian private broadcasting history. From near-poverty he rose to be one of the world's richest men through his extraordinary newspaper and broadcasting acquisitions in Canada and the United Kingdom. Under him some outstanding broadcasters gained their start: Jack Kent Cooke, Jack Davidson, Tom Darling, and others as well. He achieved outstanding prowess in the media, resulting in his best-known coup, his purchase of *The Scotsman* newspaper in Glasgow, *The Sunday Times* of London, and, subsequently, the *Times of London*. He owned half a dozen radio stations outside of Canada and a dozen TV stations, including Scottish Television of Glasgow. He ultimately became The First Baron Thomson of Fleet.

Founder of *The Sunday Herald* in St. John's, Newfoundland, in 1946, Geoff Stirling became a tycoon of Canadian radio after launching CJON, the city's second commercial station in 1949. In 1955, he started the first St. John's television station, CJON-TV. From there he headed west, founding CKGM, Montreal; CKWW, Windsor, Ont.; CKPM, Ottawa; CHOM-FM, Montreal; CJOM-FM, Windsor; and KSWW, Arizona. More Newfoundland stations followed: CJCN-TV, Grand Falls; CJOX-TV, Grand Bank; CJCR-TV, Gander; and CJWN-TV Corner Brook—all between 1959 and 1974. He set up the OZ FM Radio Network throughout Newfoundland in 1977. CJON-TV is reputed to have been the first twenty-four-hour TV station in North America. Honoured by the New York Writers Association, the California Film Board, and the New Delhi Film Board, Geoff Stirling has raised millions for Newfoundland charities through his community work.

While the television industry was struggling to cope with the distance limitations of direct broadcasting from studios, Ken Soble already was far ahead in his thinking. Why not send television's signals to a satellite orbiting the earth, and even relay them around the world? In the early 1960s, himself the head of CHML Radio and co-founder CHCH-TV in Hamilton, Ontario, Soble suggested to the Board of Broadcast Governors that he be allowed to set up such a system to feed a ninety-seven-station TV network across Canada. He died in 1966 at the age of fifty-five, before he could actually apply for a licence. Soble began his career in Toronto as a sixteen-year-old worker in radio and within a few years was the host of his own "Ken Soble's Amateur Hour." He also headed a company to produce radio shows. In 1936 he became manager and, subsequently, owner of CHML.

If anyone became a legend in his own time it was Jack Kent Cooke—originally from Hamilton, Ontario, and eventually a near-billionaire broadcasting and sports overlord in the U.S. His story is told elsewhere in this book. After his elevation from encyclopedia salesman to manager of CJCS, Stratford, Ontario, by Roy Thomson in 1937, he became owner of CKEY, Toronto, and *Liberty* magazine. A millionaire by age thirty, he made a bid for the remaining Toronto TV licence in 1960. Failing to win, he moved to Los Angeles, where he acquired radio and TV stations and major sports teams. Later, he moved to Washington, bought the Washington Redskins football team, built several sports complexes, and entered real estate, buying the prestigious Chrysler Building in New York City.

A career broadcaster from Regina, where he did his first radio stint in 1932 as a high-schooler, Lyman Potts later had a fulltime on-air career on CKCK, Regina. In 1940, he became production manager at CKOC, Hamilton, Ontario, and in 1956, as manager, put CKSL, London, Ontario, on the air. Two years later he assisted Arthur Dupont, owner of CJAD in Montreal, in obtaining a TV licence there. After Standard Radio bought CJAD in 1961, Lyman was appointed manager of CJAD's sister station, CJFM-FM, which went on the air in 1962. In 1963 he was appointed assistant to Standard President W.C. Thornton "Winks" Cran in Toronto. He was president of Standard Broadcasting Corporation in London, England from 1970 until 1974, helping applicants for radio licences in the U.K. His brainchild was the Canadian Talent Library (CTL), a non-profit recording company which recorded the music of Canadian performers. CTL had produced two hundred and fifty stereo albums by 1981.

Standard Broadcasting, Inc., was already a highly successful radio company even before "Winks" Cran took over as president in 1958. Its sole radio station was CFRB, Toronto. Cran, who came from British Decca Records, almost immediately began CFRB's rigorous preparations to bid for Toronto's only remaining private television licence. After the Board of Broadcast Governors awarded the licence to the Toronto Telegram newspaper, Cran then expanded Standard's other activities, founding CFRB-FM, which became CKFM; Standard Broadcast Sales; the Canadian Talent Library; acquired CJAD, Montreal; and moved CFRB to brand new quarters.

From crystal set listener to co-founder of the CTV network, with careers in radio broadcasting and ownership, an ad agency, radio equipment sales manager, self made millionaire, and gentleman farmer—all these and more were the story of Spencer "Spence" Caldwell. The Winnipeger who built his own crystal set as a boy became Marconi's western sales manager, was instrumental in the founding of CKWX Vancouver, and was its general manager. In 1944, the CBC hired him to manage its new Dominion Radio Network. After managing All-Canada Radio's syndicated program division, five years later he opened S.W. Caldwell Ltd. to distribute programs and broadcasting equipment. Spence's biggest bid was as one of the nine applicants for the last Toronto TV broadcast licence in 1960. After losing, he and partner Gordon Keeble founded a coast-to-coast network of private TV stations, which became CTV.

Nicknamed "Mr. Broadcasting" for his lifelong involvements in broadcasting, sales, music, and management, Ernie Bushnell was one of the best-known figures in Canadian radio, and later TV, for over sixty years. He started as a singer in 1921, founded an ad agency, became a radio announcer on CKRC, Toronto, then manager of CFRB. In 1933 the Canadian Radio Broadcasting Commission drafted him to set up its western networks. He became CRBC's Director of English Language Programming, and before World War II travelled Europe for the new CBC, assisting the BBC to organize its North American shortwave service. As CBC director of programs in 1944, "Bush" originated and produced most CBC network programs from 1933 to 1952. In 1958, he became vice president of the CBC. A year later, he resigned to set up the private Ottawa TV station, CJOH-TV, opened in 1961.

Even as a boy, Carl Pollock was a radio man. He and a chum built a small transmitter in the 1920s and broadcast music from a phonograph to neighbourhood crystal set listeners. His father, a pioneer maker of phonographs himself, later set him up in a better upcoming business, Grimes Radio. Carl, having completed study under a Rhodes scholarship, launched into research on radio technology. In 1940, he applied for an AM radio licence. With a war underway, there was no chance of that, but in 1947 he succeeded in getting an FM licence for Kitchener–Waterloo, Ontario. The lack of enough FM receivers caused a shutdown of CFCA-FM. Turning to TV in 1953, he and Ken Soble of CHML, Hamilton, combined to pry loose two TV licences assigned to Toronto, resulting in the opening of CKCO-TV, Kitchener–Waterloo, and CHCH-TV, Hamilton in 1954. Carl Pollock was by then president of Central Ontario Television, Ltd.

When Allan Waters bought CHUM Radio in 1954, there was a revolution to come in Toronto broadcasting. The first AM radio station to be founded in nine years when some war veterans got a licence in 1945, CHUM was a small station. Allan Waters, an advertising executive, changed that. He updated the broadcasting format, boosted the staff, moved the transmitter, got an increase in the station's power to fifty kilowatts, and began a unique promotional campaign. The station that had been on the air only from sunrise to sunset now went around-the-clock. Waters adopted the new "Top 40" style for CHUM, and virtually pushed rock and roll into dominance among teenagers. He expanded what became CHUM Ltd. into other broadcasting outlets across Canada, and acquired TV Muzak. CHUM-FM was launched. Allan Waters also set up cultural and charitable organizations, and opened recording opportunities for Canadian music performers through Factor/Music Action Canada.

The industrial slums of Glasgow produced a personality who, decades later, set radio listeners on their ears in his adopted city of Vancouver. Jack Webster at an early age worked for many newspapers in Glasgow and London. In the British army during World War II, he served in the Sudan Defence Force, rising to lieutenant colonel. He returned to British newspapers, becoming *London's Daily Graphic's* night editor. In 1947 he was hired by the *Vancouver Sun* to cover the labour beat, championing "the little guy" in hard-hitting reporting style. He joined Vancouver's CJOR in 1953 to do two talk shows, later switching to CKNW, New Westminster. He also appeared on the CBC's "This Hour Has Seven Days" and on "Front Page Challenge." In 1972 he returned to CJOR but left to join BCTV as "Webster," his greatest success.

Called "Bake" by everybody who knew him, Bill Baker was the prototype of the pioneer radio control operator, producer, and chief technical whiz. After working for Rogers Majestic, the radio and vacuum tube maker, in the 1920s, Bake was assigned to run controls for the on-air debut of CFRB in 1927. He and Wes McKnight were the station's first employees. Over the next fifty years or so, Bake supervised the growing staff of technical operators as RB expanded, especially in remote sports broadcasts, which Wes announced. In 1965, under station chief engineer Clive Eastwood, Bake and his technicians successfully transferred CFRB and CFRB-FM from Toronto's Bloor Street to slick new quarters at St. Clair Avenue and Yonge Streets.

the march of technology

The public's appetite for timely information burgeoned, and radio and TV responded even faster. New technology allowed better taping and reporting equipment. Tape recorders shrank from suitcase size to hardly more than that of a wallet, making the radio reporter more mobile. Devices appeared that allowed a radio reporter to convert a phone booth telephone mouthpiece into a higher-quality mike by removing it and substituting a better one that gave broadcast quality to the telephoned voice report.

The ultimate great leap forward came years later with communication by satellite, making the reporter sound like he or she was right in the room. Ken Soble's augury of things-to-come was being realized. The audio quality became superb. A reporter in Timbuktu sounded like he/she was next door.

Old hands bemoaned the loss of the raspier on-the-spot report with its crude sound and ring of urgency and immediacy. The new fidelity was too good to sound like a report was coming from a remote location. It wasn't convincing, they felt. But radio's unswerving drive for on-air technical perfection had to be recognized as one more great step for humankind.

Audibility was becoming a must for audiences. Live reporting was no longer thought of as the superhuman effort it had seemed in older days, when reporting from remote places appeared fraught with difficulty and extreme hardship for the reporter. That was not always the case. However, it was part of radio's illusion, and the pops and cracks and rush of static helped reinforce that impression. But public ears were becoming more sensitive to the harsh sounds of remote reporting.

In the mid-1960s, radio was emerging as a major source of public information, and listening to it needed a high fidelity sound, especially when heard inside a car filled with engine and fan roar, or in other noisy spots. The traffic-bound commuter wanted to hear about the highway jams up ahead before he/she got there, and was coming to rely on the station with the clearest sound.

But other aspects of life were to be heard on radio, too. CFRB's popular Betty Kennedy, for example, in the mid-afternoon brought out revealing facts from interview subjects in all walks of life in her casual, probing way. It helped car passengers pass the time spent in traffic jams informatively, and also learn about some significant person or process. Bill Deegan coolly talked up music and commuting tips the rest of the afternoon. Music had to be heard as though in a concert hall. Gordon Sinclair provided bombast on his ten-to-six newscast. If commuters were still stuck in traffic by six-thirty, they were brought up to date on the news by Jack Dennett, in his booming baritone and authoritative style of delivery. Especially with the onset of FM, clear sound was in demand.

And then there was the road-roving traffic reporter. Before Eddie Luther took to the air in the helicopter, it was part of my homebound duty to travel segments of Highway 401 in particular and report on accidents. Sometimes I'd go out in one of the CFRB Traffic station wagons, heavily loaded with radio transmitting equipment connecting to the station. But for awhile, ages before cellphones appeared, a wireless telephone transmitter took up a large part of my car's trunk, with the phone itself perched on top of the dashboard.

This, Bill Hutton theorized, would make it possible for me to phone in instant reports the minute I spotted some accident, or other news happening on my way to or from work, so that not a minute of my time would be wasted. To do reports I would have to get off the highway and up onto an overpass, where the nearest line-of-sight Bell receiving tower could pick up my transmission more clearly and relay it to CFRB's control room and then onto the air.

Often, at crucial moments when instant reporting of a spectacular accident was paramount for homebound drivers to know about, I would pick up the phone — shared with several other Bell customers — and hear talk going on between, for example, a man and wife who kept me speechless about some traffic crisis that faced hundreds, perhaps thousands.

I'd hear a male voice: "But honey, I'll only be a couple of hours and then I'll be home. This meeting's important." This would be followed by a slow, low-pitched female voice muttering doubting negatives, such as "Oh yeah, I've heard that one." After several attempts at getting control of the phone it would be too late to bother reporting the accident and alerting thousands of drivers to take another route to avoid it. They just drove into the mess unawares. Domestic shenanigans took precedence. Thankfully, the CFRB brass decided to go for hiring a helicopter, and putting Eddie Luther in it to tell the traffic story morning and evening, thus pioneering an eye-in-the-sky service that other stations eventually adopted. Although I, too, was sometimes up in the air in the chopper, I was back on real ground-level news reporting.

While I had the car phone, however, things happened that made it pay off. There was, for example, the night I was on my way home after preparing the eleven p.m. "World Tonight" newscast for Gordon Cooke, when the phone rang. It was Gordon. He was actually in the studio doing the newscast but he had a taped report on the air and could talk. In his super-calm way he said there was a tremendous fire at the Canadian National Exhibition grounds. Could I do a side trip and see what it was about? Okay, I said, despite being ten miles away from the CNE.

I got there, though, even before the fire department. The Ex being parkland near Lake Ontario, and far from residences or places of work, apparently nobody had noticed the fire in its early stages except for our mysterious tipster who'd phoned Gord Cooke. How did he/she know about it? It was by now a spectacular sight. I drove onto a wide grassy area of the CNE grounds, and there before me, a hundred yards away, reared up the inferno. The Ex's old Manufacturer's Building was spectacularly on fire.

I must have been the only witness at that point. There were no other reporters in sight. I got the phone going and Gord's control room operator put me directly on the air. For about ten minutes I rattled off an eyewitness report. The building had a row of arches ranging across front and side, and through these arches I could see a broiling, roaring hell of flames rapidly eating up the building. As far as I knew, no one was in it. Little by little it collapsed in horrifying style.

I still seemed to be the only one around. Then the fire trucks arrived. Maybe my report was heard by the firemen; who knows? Or perhaps our tipster was busy again. The firemen got to work pouring water on the blaze. Once they'd doused it enough that it looked manageable, there was little left for me to say, so I signed off. Also, it was late — well past midnight. I was tired. Not much was said about our scoop the next day. Probably none of the brass even heard it. But there was satisfaction between Gord Cooke and me that we'd done right by the audience and radio in instantly delivering exclusive news *live* on the air right from the scene.

CFRB came close to another major scoop when the 1963 federal election duel between Prime Minister John Diefenbaker and Liberal leader Lester B. Pearson was at its height. Pearson's crew alerted the news media that their leader would hold a press conference at Toronto International Airport — later to be named after Mr. Pearson himself. The Leader of Her Majesty's Loyal Opposition would pass through on one leg of his election campaign. There was nothing for it but that Toronto's leading radio station be there, along with all the others.

I put together recording equipment and a long microphone cable for the job.

At the time, we had the young Canadian track star, Bruce Kidd, working for us in the newsroom as a summertime newsroom assistant between his University of Toronto years. A quiet, friendly young man, Bruce was willing to do whatever he could to help. Hutton assigned him to go with me to the Pearson press conference.

Together we lugged the heavy suitcase tape recorder and a great long coil of heavy cable out to a CFRB News station wagon and headed for the airport. On the way, I recalled that Prime Minister John Diefenbaker had had some caustic things to say about Lester Pearson in an appearance somewhere the night before. That's what I'd ask Pearson about. We pulled into the parking area and lugged the equipment into a long narrow room packed with news people. A long narrow table was set up, with chairs ranged along it. Pearson hadn't shown up yet.

There was no room left on the table to put the heavy tape recorder, so we put it on the floor behind the row of chairs, and hooked up the mike to the cable and the cable to the recorder. It was when I was telling Bruce to stay with the recorder while I handled the mike that I spotted the gaffe any radioman dreads: I'd forgotten recording tape. The predicament was like a plumber forgetting his wrenches, or a photographer forgetting his film. But I had an idea. My own house was within about five miles of the airport. I could dash there in the station wagon and get a reel of my own audiotape from home and dash back.

I left Bruce with the equipment, and dashed off. It seemed to take no time at all to pick up a tape and get back. The tape actually was already recorded with some of my favourite music, but I was willing to sacrifice it. As I slapped the reel on the recorder, Lester Pearson and his crew entered.

I took the mike on its long cable up to where I could hold it right under Pearson's mouth and get most of what he said during the press conference. I'd met him before and he sort of recognized me, so he didn't object. Everything was going well. As the press conference broke up, Pearson was delayed just inside the exit door by a reporter or two with more questions.

Bruce and I were ready to pack up when I decided to pop the question about Diefenbaker. I took the mike again to Pearson's mouth while standing just partly behind him. When I said, "Mr. Pearson, Mr. Diefenbaker last night called you 'a weak-minded poltroon.' What do you say to that?" Pearson swung around, stung, and let out a blast against the prime minister: "If I'm a poltroon then he's a fully-qualified Ringling Brothers' Circus baboon!"

I was the only one with a mike in front of him to record this memorable retort. Instantly, he was off like a shot to barnstorm away on his waiting airplane. On the tape, I added a one-minute voice report on the event for the newscast, allowing for the insert of Mr. Pearson's retort.

I knew I'd scooped the other radio news people with the Pearson reaction. Hurriedly, we collected the tape recorder and the cable and headed out of the airport. Back at the station, I gave the tape to an operator to edit, and since it was the end of my working shift, I went home. There I tuned-in to RB for the first newscast I thought the Pearson tape might be used on. There was no mention of it. Other quotes from Pearson off the wire were used, but not the tape. As I wondered what was happening, my phone rang. It was Jack Dawson, the formidable CFRB station manager.

"Gil," he said, in his usual quiet but dangerous style, "what happened to that tape you made at the airport? It sounds all goofed up."

And he had an operator play the tape over the phone. It certainly was goofed up. The voices on it from the press conference, including Pearson's and mine, made a warbling sound that turned us all into Disney Donald Ducks. At first I was puzzled. Then I realized what must have happened. As our tape recorder was sitting on the floor right behind a chair, some reporter must have got up and unthinkingly shoved back the chair without looking — or on purpose.

As my tape was on a large seven-inch reel instead of the usual four-incher, it poked over the tape recorder's edge just enough for it to scrape against a chair leg but keep rolling, though wriggling. As it rolled, the wiggling tape recorded the voices as a warble.

Trying to explain that over the phone to a station manager obviously just able to control his anger at this goof was beyond the pale. But Jack didn't explode. He muttered something and hung up. What bothered me most was that my scoop of a monumental political figure's immortal, scathing reply to another monumental political figure's insult never was preserved. If it had been, CFRB would have caught nation-wide attention with an exclusive newsmaking reaction, typical of the bitter 1963 election campaign, by the next prime minister of Canada.

In the fullness of time, as former Ontario Premier Leslie Frost liked to say, there were more encounters, however brief, with the great ones of the era. Hubert Humphrey, still a U.S. senator and not yet vice-president of the United States, came to town, and I received a phone tip. He was staying at the Park Plaza Hotel. I knocked on his door, and it was immediately opened by Hubert himself. There stood the newsmaking senator who was soon to become U.S. President Lyndon Johnson's vice-president, and who himself would later run for the presidency, but fail. No other reporters turned up. I was surprised that he had no covey of PR people to handle him. But, let's face it, there were no votes in Canada for him.

He welcomed me into the suite, shaking hands as though I were indeed a prospective voter — probably from ingrained habit — and we sat down together on a chesterfield where I could put my tape recorder on the low table in front of it. First, though, he had to tell me about the minor car accident he'd been in the night before while being driven into Toronto. He showed his bandaged shin where he'd been hurt. From his folksy-homey style, I could see how his simple Minnesota small-town charm could work on voters.

That out of the way, he said he was in Toronto on behalf of a Red Cross fundraising campaign, and described the need for more cash for the Red Cross to do its job. His reply went on for several minutes, leaving no more to be asked. Then, just like long-lost friends after a happy reunion, we shook hands heartily again, I packed up the taping equipment, he wished me well, and saw me to the door. I don't think the tape ever got played on the air. The Humphrey presence was not yet spectacular enough to be worthy of broadcast to Canadians. But CFRB News did actually tape the future U.S. vice-president first. The tape hasn't survived.

The encounter with Diefenbaker man George Hees was different. Here was a man in self-conflict and in the news, important in the process that later brought about the Chief's downfall. I was working in the RB newsroom on a Sunday morning when I got a phone call from our Ottawa bureau. There'd been a big blowup between Prime Minister Dief and some of his cabinet in Ottawa. Hees, the minister of trade at the time, and a reluctant Diefenbaker cabinet minister, had finally given the Chief a blast, quit the cabinet on the spot, and was last seen heading home to Toronto.

I found out that home was in a high-rise not far from CFRB's Bloor Street studios. Grabbing a tape recorder and a CFRB cruiser, I headed along Bloor and arrived on Hees' floor just as the newly-resigned cabinet heavyweight reached his own apartment door. There he was, in snowless Toronto, in regulation Ottawa garb: a ton of overcoat and a Persian lamb fur hat. He was obviously not pleased to see me. His eyes were bloodshot and teary. He was plainly very upset. However, still the politician, he knew the value of a few words over CFRB.

A big man, a noted war hero, Hees was plainly surprised that one of my breed would be so promptly on his doorstep after what must have been a catastrophic showdown between him and Dief. He had only a few words to mutter into my mike. He gave away nothing about what had brought on the cabinet revolt against the fiery PM. Then he disappeared into his apartment, and that was that. Hees hadn't said much, but the incidental fact that a radio station and not a newspaper had intercepted him so soon after the event was a plus for radio's ability as a news deliverer, for those who cared. It also showed radio as an instant-reporting medium to be reckoned with.

I myself had faced the wrath of Dief the Chief at a 1963 electioneering press conference as he stood at a microphone while I and other radio newsmen crouched low to the floor with upheld mikes. As I was the handiest, I suppose, the PM fixed his terrible stare on me, pointing a finger directly at me as an apparently ghastly example of all irresponsible news media, and roared a general condemnation. I ducked, grinned weakly, and Dief switched to another fire-eating diatribe. I don't think it went over well with the rest in the crowded room. In the end, Dief lost the election to Mike Pearson.

As Ontario Press Gallery President, I sat in a King Edward Hotel suite with Ontario Premier John Robarts before the two of us went down to the landmark Toronto hotel's Crystal Ballroom for dinner — he to sit in the front row beside the Lieutenant-Governor at the after-dinner "Press Gallery Follies" show, I to start the show as master of ceremonies. Before we left, the premier had to make a phone call, but even with his glasses on he couldn't read the phone dial. "Gil," he said, "I can't see the numbers. Can you dial this for me?" He'd dropped his guard against news media. Mine was the surrogate dialing finger for the Premier of Ontario. I never knew who the personage was he was calling.

That was a revelation: "Premier of Ontario Too Blind To Dial Phone" might well have been the heading I would have dutifully written on the story if I'd still been on the *Toronto Star* desk. However, in my temporarily exalted position, I couldn't breach confidentiality. I dialed the number. Later, onstage, when I was introducing the "Follies" — a crazy costumed lampooning by press gallery members of every main government official from the Lieutenant-Governor to the Premier and his cabinet down to the Leader of Her Majesty's Official Opposition in front of the victims themselves and about two hundred civil servants — I forgot to call for The Toast to The Queen.

Most of the audience didn't notice. But some protocol hound did. After all, we had the Queen's Ontario representative as our chief guest. We backtracked on the proceedings, raised glasses, and sang "God Save The Queen" before we went on. From the calibre of the show, with some of the journalistic performers and most of the audience well fortified for the occasion, Her Majesty would probably, if she had known, have been happier if we hadn't mentioned her.

There were other brief moments of transient glory. Bill Kinmond, Robarts' press secretary, one late evening called down to me from a Queen's Park mezzanine balcony to hang around for awhile, just as I was heading for the door after a long day at the legislature. It turned out that he had a scoop for me: The premier had just picked the successor to Kelso Roberts, the Attorney General dethroned over the gambling scandal. I was the last newsman around. I had the exclusive news on that evening's "World Tonight" that Arthur Wishart, an obscure back-bencher, was now Ontario's new Attorney General, the enforcer of the law and the highest in cabinet.

Now it remained to be seen how Wishart would swing the axe on everybody connected with the gambling scandal, inside and outside of politics. It was Robarts' first shakeup of his cabinet. And radio had it first.

An *almost* non-publicized event brought together Premier Robarts and Quebec Premier Jean Lesage in 1963, through the good offices of the Ontario Press Gallery and the Quebec Press Gallery. A suggestion for a three-day official visit to Quebec City and the Quebec Press Gallery came to the Queen's Park Gallery from some unidentified source. Ostensibly, it was simply a professional "goodwill" visit between the two galleries, with courtesy attendance by the two premiers, and, incidentally, some top cabinet ministers and some senior bureaucrats. Radio, both from Quebec and Ontario, was well represented.

In reality, as I discovered later, it was really a disguised informal get-together by the premiers, genuinely concerned by the imminent birth of the Quebec separatist movement. While our large press gallery delegation from Ontario went on Quebec City tours and partied with our Quebec counterparts, Robarts and Lesage met privately. What they actually discussed was strictly confidential. The two press galleries discreetly ignored the content of their talks, as the premiers' aides cautioned that publicizing them might throw a spotlight upon the yet publicly-unspoken, and delicate, topic of national unity.

We and our wives were formally and elaborately welcomed in the Quebec Legislative Chamber. Later, touring notable Quebec City sites, we passed the bare pedestal of the statue of General Thomas Wolfe, victor at the Plains of Abraham in 1759, which drove Louis XVI's troops back to France. A few weeks before, separatist terrorists of the FLQ had blown General Wolfe off his base with plastic explosives. He was never put back together on it.

We were feted at lunch at the Quebec Lieutenant-Governor's residence outside Quebec City, the incredibly beautiful eighteenth century Bois de Boulonge. A month or two later, the rambling mansion mysteriously burned down in a fire whose origin was never determined. Our gracious host the Lieutenant-Governor and his daughter perished. Otherwise, we saw no sign of a Quebec separatist revolt.

At lunches and dinners where we resided at the Château Frontenac during the three-day visit, attended also by a young, then-Liberal Quebec cabinet minister named René Lévesque and possibly also by an equally young Pierre Elliott Trudeau — a working journalist — we heard lengthy speeches by Quebec politicians and philosophical luminaries. As they were all in French, no English, and this was long before the era of Pierre Trudeau's bilingualism policies, few of us understood them.

All in all, looking back, we'd had a part in what might have developed into a thwarting of Quebec separatism through a combined effort by Quebec and Ontario, a movement at the time not of any public interest. Had Robarts and Lesage managed to formulate some quiet anti-separatist moves, the troubles fomented later by the terrorist FLQ and quashed by Prime Minister Trudeau's War Measures Act might never have taken place.

We Ontario Press Gallery members, who'd unwittingly provided the smokescreen for the two premiers' talks, thoroughly enjoyed our sojourn in Quebec City, though we didn't have a clue what it was really all about.

out of the frying pan

Working in radio was somewhat like working for the railroad: while everybody else is on holiday, the radio guy is working. I'd already decided in March of 1968 that it was time to move into a line of work that didn't demand constant alertness to constant deadlines, outrageous hours (I was now coming in at five-thirty a.m. to launch SBN's first daily transmission), short vacations, and no prospect of pay increases. One job I didn't ask for was as guardian against unwanted visitors to the radio station.

Besides fielding nuisance and often obnoxious phone calls from unnamed people who had a gripe about something said over the air (most such calls were piped through from the station switchboard to the newsroom to be dealt with), occasionally a belligerent, often drunk, individual would turn up at the front desk, insisting on fighting everybody from Wes McKnight, to Gord Sinclair, to Wally Crouter, to Bill Baker — this wasn't uncommon. Radio stations and newspapers seemed to attract such pests.

Over decades, Baker had to manhandle a few of them back down the stairs at the old CFRB Bloor Street quarters, and shove them out the front door. I unluckily got the job of ejecting one such visitor who was bothering the receptionist, which I achieved mostly by talk and a strong hand on the man's arm. Afterwards, Baker nodded knowingly when I mentioned the incident. He'd had many a fistfight.

"Just don't hit them in the face," he said, recalling one such battle. "That only gets them mad and they fight more. Hit 'em in the guts."

I remembered this kindly advice, daunting as it was, when my next encounter came. We'd just moved to the new St. Clair and Yonge Procter and Gamble building. The second-floor station was normally well-secured, with the receptionist safe behind a plate-glass window and the door into the studios equipped with an electric lock that she could control. She could see everything on the stair landing and down the stairs through her window. She had her switchboard phone to call for help, if needed.

However, at six a.m. she wasn't there, and the only people in the station besides me were Butch Harrison, a veteran operator; Wally Crouter, the early morning man; and his operator, Bev Edwards. An overnight newscaster was just finishing up his shift. The rest of the station was deserted. It also was heavily soundproofed. No noise from the corridors could penetrate into Bev's control room, Crouter's studio, or the newsroom.

Having sent the day's first Standard Broadcast News feed across the country, I headed down the corridor to the reception area to get the morning newspaper. This was always left just outside the door. I made the mistake of not looking out the receptionist's window first to see if the landing at the top of the stairs just outside the door was vacant. It wasn't. I found that out when I slipped the lock, opened the door, and started to stoop down for the paper. It was then that a body lunged at me through the half-opened door.

I grabbed it before it got all the way through and jammed the door against it as well. But the man was already half in, and wasn't going to quit. He was a stocky young fellow with a shaved head and wearing the white canvas jacket and pants of the kind that inmates conventionally wore in mental hospitals. He was whimpering, but pushing with great strength to get past me. I couldn't allow that to happen. If he got away inside the station and burst in on Wally Crouter there would have been a commotion broadcast to all of CFRB's vast morning audience and maybe even a murder on the air. The scapegoat would have to be me.

I held the mystery visitor jammed in the doorway as he whimpered, over and over, something about wanting his mother. This was a predicament that seemed to have no solution. I couldn't let him go, nor could I hold out indefinitely against this husky intruder until some of the staff turned up and called police. That was still about two hours off. And, remembering Baker's advice, I couldn't sock him. We just went on heaving against each other, neither winning nor giving in. I yelled down the corridor, but because of the superb soundproofing, nobody could hear.

After perhaps twenty strenuous minutes of pushing and shoving, I was delighted to see a huge policeman appear at the top of the stairs outside the door. I expected him to grab my gladiator and cuff him, as they say on TV cop shows. But my cop just said: "Well, Marty, what have you been up to this time?"

He stood there watching the shoving match for several minutes. Finally, I yelled: "For god's sake, get him off me! Grab him!"

At that, the cop, now joined by another one, grabbed Marty, and the two of them slammed him down on the marble floor and did "cuff him." My body sprang back from the position it had held for over twenty minutes. I hurt all over. They carted Marty off down the stairs. Watching all this was a small taxi driver, standing in a corner outside the door.

It seems that Marty got into his cab outside the Queen Street West mental facility and told him to take him to CFRB to see his mother. Marty told him to wait until he got cab fare from his mother (at six o'clock in the morning!). When he didn't show up for nearly half an hour the cabbie called police, who came and, casually, did their duty.

Furious at having been caught in such a situation, I told him he'd have to come back later for his money but, anyway, I didn't know who Marty's mother was. I took the newspaper and stalked off back down the corridor to the newsroom. Everything there was normal. In the broadcast studio, Crouter was chattering on the air, Bev was getting the next commercial tape ready, Butch Harrison was getting ready to leave, and Torben Wittrup was typing up the first of his day's newscasts. I barged in on Wally Crouter while a record was on the air, and he and Bev Edwards were verbally jousting over the talkback.

When, no doubt sounding overwrought, I told Crouter how I'd just got through saving us all from a wild man at the front door, he listened for a moment, then wisecracked about the next record through the talkback to Bev. I left. Somehow, I felt, I must be transparent or not audible. Back in the newsroom, I set about typing up a story off one of the phoned-in tapes for the next newscast.

I let the whole thing go. It turned out that Marty's mother did indeed work at the station as an office clerk, but she wasn't due in until eight-thirty. In my own eyes only, I remained the unacclaimed martyr who'd saved CFRB from a morning of mayhem. And I'd lived up to the Baker credo: I hadn't hit him in the face.

Christmas of 1965 was approaching, and the CFRB crew was well settled into Canada's most modern radio station — being the latest built — at Toronto's St. Clair Avenue and Yonge Street. In the brand new, shiny, clean working areas and with more space than at 37 Bloor Street West, the atmosphere seemed less tense than it had been at times in the cramped old building. By adequate allotment of recording and duplicating equipment, old frictions between the operating staff and the news staff over who got to use tape-editing equipment first were gone.

Now the newsroom had two big floor model Ampex tape recorders taking twelve-inch tape reels, interconnected for quick dubbing of news clips, with two of the new "cartridge" recorders/players (similar to non-professional tape cassettes then coming into public use) tied in. Now we could re-record snippets of long reports more quickly to produce news clips. In spite of time pressures, one could settle into being comfortable, even almost complacent, in the new surroundings.

That may have been why Bill Hutton informed me of my next major assignment. It was all set up, he said. There was going to be, a week before Christmas, a flight by the Royal Canadian Air Force (the RCAF still existed as such) to the sub-Arctic, and I was to be on it. Now, I'd served in the tropics of Australia during the war, not all that unhappy with the heat, and I'd never had much love for cold weather. Visiting the sub-Arctic in winter? Just before Christmas? That didn't really have much appeal, especially to a family man. But that was radio. It demanded all of you, not just part.

I wouldn't have balked at a winter assignment to, say, a Caribbean island, but Bill had already cornered that himself, spending some time in Barbados covering the HARP project (the High Altitude Research Project of Professor Gerald Bull of McGill University). He didn't seem willing to seek out sub-Arctic work. So there was nothing else for me to do but carry the CFRB News flag to the wilds of Hudson Bay myself. I muttered that I was sure it would be interesting. It was to be, in fact, a unique expedition. Not too many private radio broadcasters had likely deliberately ventured into the sub-Arctic in search of a story. Once again, radio would do pioneering work.

This was a mission the RCAF undertook every year, flying presents and Christmas parties to the Innu on the Quebec side of Hudson Bay, and then over to the Cree on the Ontario side, both tiny settlements just where James Bay opens into Hudson Bay. The nomadic Innu came as far south in Quebec as Great Whale River, but still over seven hundred and fifty miles north of Montreal. The Cree at Winisk in Polar Bear Park on the Ontario side, even farther from civilization at over eight hundred miles north of Toronto, hunted and trapped in the tundra for miles around. Here we need some geographical orientation:

Winisk was, and is, a Cree village just northeast of the joining of the Shamattawa and Winisk Rivers, a desolate spot in winter. The Pine Tree Early Warning Radar Line, the most southerly of the chains of radar stations stretched across northern Canada to warn of incoming Soviet missiles, if any, had stations in both Winisk and Great Whale. The RCAF had air stations there, too. They would be hosts for the Toronto news people.

Many Southern Ontario media, including the *Toronto Star*, the

Telegram, and the *Globe and Mail* were sending reporters and photographers. There were some from out-of-town papers, one or two radio stations, and several magazines. In 1965, it was apparently still too early in their mobile capability for the Toronto TV stations to send crews along, although there was one from CHCH-TV in Hamilton. The media complement came to about twenty men and women.

At Downsview RCAF station, we were briefed on what to expect. We'd be given air force parkas and thermal pants to wear over our suits or whatever we chose to travel in. We could expect daytime temperatures of about 30°F below zero, (no Celsius yet but it would be about the same) and only the gods knew what it would be like overnight. We'd first fly to Montreal to pick up some entertainers and then take off for Great Whale. Altogether, we'd be gone three days.

The plane we'd fly in was a North Star, a four-engined aircraft, originally one of a civilian fleet bought by Ontario Premier George Drew's government in 1946 to fly immigrants directly from Britain to Ontario. They might have been well-equipped then for comfortable flight, but the military had since adapted them for paratroopers. That meant that, to lighten the plane's weight to carry about fifty muscle-bound soldiers and gear, they'd stripped out all sound and cold insulation, removed all comfortable seats, and slung canvas-and-pipe seating along both inner walls of the fuselage.

When the four engines started up, all talk ceased. The roar was so deafening that you could feel your eardrums rattle. No hearing-protectors were issued. Over the next three days — mostly in the air — we would endure an incredible assault on our hearing. To communicate, we had to put our mouths up to another person's ear and shout. Among this group of professional communicators, there wouldn't be much communication.

Along the middle aisle, huge packing crates were strapped down, presumably holding gifts for the Innu and Cree. Later, after Montreal, there would also be big boxes of musical instruments, costumes, and so on. That, plus the fact that in the low canvas seating, with a pipe cutting into the undersides of our thighs, and all crammed together with no such convenience as an armrest between us, the flight wasn't going to be exactly a fun time. On the other hand, the presence of so many bodies helped keep us reasonably warm.

christmas in the sub-arctic

We landed at Dorval airport in Montreal, already all shaken up after only three hours in the North Star. We piled out for a coffee break. We got back to the plane to find a major loading-up going on: more boxes and additional humanity totalling several young men and half a dozen or so girls. These were the members of two Montreal rock bands and a small dancing group. Their boxes of instruments were added to the crates already in the aisle. From then on we couldn't see each other across the aisle. With the engines silent, the plane was full of cheerful chatter, mostly in French. That ended when the first engine was fired up.

It was still daylight when we took off, heading north. In no time at all, it seemed, it was dark. Every now and then as the plane droned on, there would be a single dot of light in the vast spread of northern Quebec forest forever below us. In the plane, there were only dim interior lights to see by. Reading anything created eye strain. Chatting with one's seat-mate on either side was impossible. Hours were to go by before we reached Great Whale River. A snack and some canned pop were passed around. I thought to myself that this had better be a worthwhile radio event.

The young band members looked anything but equipped for the cold. They wore only thin clothing and light windbreakers. The plane's interior was above freezing, but how would they stand the cold of Hudson Bay? I managed to shout that at one of them. Oh, they'd been on this trip before in the last couple of years, and they didn't worry about it, he shouted back. This, though, was to be the last Christmas trip by the air force. The RCAF wouldn't exist after the new year, but would become part of the new Canadian Armed Forces. Frills such as these trips would be put on hold.

As it turned out, it was also to be the last trip for this last North Star, by then over twenty years old. Contemplating this while cruising over the wilderness below, in winter, was not an especially comforting thought. Somehow, despite what might be our precarious position, and over the engines' roar, the young Quebec veterans of this annual junket kept up their lively chatter and antics. Most of us fell asleep.

At long last, the sight of a patch of light in the black below energized everybody. It was the airfield of Great Whale River, the only civilization for hundreds of miles around, perched on the east shore of Hudson Bay. A single runway was well lit. From the air, it hardly looked big enough to let a four-engined plane land on it, but land we did.

There were some huge RCAF airplane hangars and numerous H-hut barracks buildings nearby. A crowd of small figures in brightly coloured parkas with white fur trim and tall peaked hoods with white tassels on their tips clustered at the front of one of the hangars. It looked as though Christmas elves had gathered to welcome us to Santa's Village. They were Innu. Some were singing. The arrival of our North Star was undoubtedly a major event in Great Whale River.

Snow lay all around, not as deep and even as one might have expected for the far north. In the masses of floodlights, with huge clouds of steam rising from them and the barracks huts in the extreme cold, our North Star looked huge as it was manoeuvred immediately into the hangar. It wouldn't take long for its engines to freeze up if left outside. The sound of diesel engines of tractors and other heavy vehicles would rumble all night and all day. In that cold, they were never turned off.

The Quebec kids, knowing from experience what the cold could do, had already made a beeline for the nearest building in their skimpy clothing. After much waving of hands at the little Innu at the hangar, we media types ambled in our parkas to the same building, which turned out to be a recreation centre and cafeteria. A meal at last! Inside, it was like all the barracks buildings I'd known back in the army, but at least it was warm and reasonably comfortable — and quiet.

The Pine Tree Line was operated by civilians from the Marconi company in Montreal. The living logistics were handled by the RCAF, who kept a few CF-100 fighters at the air base. So the resident group that welcomed us was not only RCAF, but civilian as well. The Marconi people spent short terms of duty at Great Whale, and then perhaps also over at Winisk, before going back to Montreal when they finished their sub-Arctic tour of duty. Another contingent would replace them.

The Pine Tree Line served its purpose into the 1980s, when, with the collapse of the Soviet Union, it went into limbo. Both it and the far north DEW (Distant Early Warning) Line became redundant. No nuclear-tipped missiles from over the Pole were expected any more. The sub-Arctic reverted to the Innu and the Cree and the business of trapping.

Another huge hangar was decorated for the Christmas concert, with a stage at one end and two long and wide sections of chairs, one side already filled by Innu, still in their parkas, but with the hoods down. On the other side of the centre aisle the seats were filled with Cree, also from the district. No words or greetings passed between them, not even glances. There was total separation.

A priest, lingering with us at the back of the hangar/hall, chuckled when I asked why there was such a clear division, and said the Innu and the Cree just didn't mix. They carried into this hangar the same tribal remoteness they held between them outside. Great Whale River was the line drawn between the two races. There was no open conflict, he said, but nothing could bring them together.

One of the young Montreal rock groups, now outfitted in showbiz garb, was thundering away with their rock music when we came in. The photographers instantly began shooting. Both the Innu and the Cree, of all ages, sat immobile in their chairs, transfixed by the rock music. I taped some of it. Who could tell? It might make good background for an RB report that I would have to do.

The concert slid into Christmas carols, one of the rock groups changing to the role of a sedate accompanying band. There were Christmas skits by some of the Montrealers, in French, and a number featuring a boogie between showbiz girls and teenage Innu, to huge applause. But for most of the time the Innu/Cree audience sat without a move, attention fixed entirely on the stage.

A gala distribution of Christmas presents to the two-part audience took place, with one of the bands blasting out rock versions of carols. At concert's end, like any audience, the native people began to leave — the Cree out one door and the Innu out another — all hugging their presents. The media and the RCAF'ers retired to the officers' mess in the recreation hut, where the bar instantly began to boom.

I got on the phone and ad libbed a report to CFRB in Toronto and an RB night operator who was startled at getting a call from a staffer away up in the sub-Arctic. He'd thought I was home in bed. Indeed, after a comradely visit to the barroom, I headed exhausted to the barracks bunk assigned to me in the familiar military style. Unlike my old army bunks, this one had sheets and pillowcases.

Next morning, the air force ground crew wheeled the North Star out of the hangar on to the tarmac, turned it on, and let it roar for half an hour or more to get the engine head warm enough for takeoff. When we taxied away, we left the parka-ed Innu waving a wistful goodbye. Their last annual RCAF Christmas visit was over. They shrank away and disappeared into the white as the North Star headed west to fly across Hudson Bay.

The bay was a vast table of ice, a symphony in white under brilliant sun and clear blue sky. As we flew over it and neared the western shore, a tall spear of smoke was rising high and perfectly straight up from a single cabin buried in snow on the edge of the icebound bay. What brave soul might be living there through a sub-Arctic winter was a thought to muse upon.

The whole scene was icy-crisp. Eventually we were circling over Winisk. Off the mouth of the Winisk River there was a blue slash of open water in the Bay, seeming all the bluer against the white landscape. The landing was similar to Great Whale's. The difference was that there were no Innu on hand. This was strictly Cree country. Everything was just as cold as at Great Whale, but there seemed to be a more visible RCAF presence here.

The base looked much like the other one, with clouds of vapour gushing aloft here and there from barracks huts and idling engines. But Winisk seemed to have more character. One of the reasons may have been that the base, as with many establishments headed by a strong personality, was commanded by Squadron Leader Kyle Watson, former leader of the RCAF's original crack aerobatics team, the Golden Hawks. Stocky and cordial, he shook hands all around in welcome as we southerners stood with the cold nipping at our exposed noses. I checked a thermometer on a building wall that read -35°F — in the sun! Still, it didn't really seem that cold. But again, the young Montrealers sped from the plane to the recreation hut in nothing flat.

As we had the day free to sightsee — although we thought we could see everything from where we stood — once we'd been settled in at the barracks, Squadron Leader Watson asked if anyone wanted to visit the Cree village a mile or so away. He would take us himself in his big caterpillar-tread snow cruiser. Six of us — reporters, photographers, and a CHCH-TV cameraman — said that would be interesting. After all, we were on news assignment and there might be something newsy over there.

We'd already seen Natives racing around in strange, ski-equipped scooters, slewing around and running them up ramps to jump off into the air. Years later, this kind of vehicle would evolve into the snowmobile and become beloved in the south. The Cree were having a winter meet featuring tricks on the snowmobiles. Watson's snow cruiser was something different. This was a big box, the size of a truck, with windows and board seats that would hold about a dozen people and driver. We — no women — climbed in. Up in front, Watson shoved the vehicle into gear and hauled on the two upright levers that substituted for a steering wheel, just like a tank. We crawled forward, the caterpillar treads clanking and squealing.

There was a pretty good road to the village, but our driver shouted out over the engine roar that he was going to take a cross-country shortcut. He cut off the road and we were instantly jolting over humps of frozen snow so rough that we were bounced off the wooden seats. The idea was to haul ourselves up off the seats by overhead straps to save our bottoms and swing like monkeys as our squadron-leader, yanking levers like a madman, sent us over the tundra at twenty miles an hour. We swung all the way, cameras and tape recorders dangling from us on their straps. Watson yelled over and over that it wasn't far now. But we could see he was chuckling mightily over this pummeling of the perfidious news media.

We came to the Winisk River. Our snow-cruiser bucked, tipped, and clanked down the bank to the frozen river, which Watson yelled was no problem: the ice here was twelve feet thick. Once on the river ice, the ride smoothed out and we could sit. Then, with the streak of open water in the bay coming into sight, Watson skewed the cruiser around to climb the other riverbank. It got part way up, then stopped, stuck in the snow, pointing skyward at a forty-five degree angle. Watson remarked that the vehicle had no reverse gear, so we'd have to get out and walk to take weight off so he could get up the bank. We climbed out.

Watson revved the engine, and gradually the snow-cruiser scuttled up and over the bank. The prankish Watson didn't stop for us, but kept his vehicle going. That didn't matter much, as we could see the village not far off. But the frigid temperature began to bite. So we hardened newsmen trudged rapidly across the tundra in our city shoes to the rows of houses.

There was a howl of huskies. We came upon the huge dogs chained to thick stakes driven in the ground on the edge of the village. I pulled up my portable tape recorder to get some of the howls as more background for my Christmas report. It was frozen stiff. So much for background colour. I tried my camera, got two or three shots off, and then the shutter froze.

The village was nothing special to see. There was a row of houses on both sides of the road that went down to the bay. What was interesting was that these aluminum-sided houses were two-storey, recently built, and much like the homes you would see in any southern suburb, except for the total absence of landscaping or trees. I hadn't thought trapping beaver would bring in that much revenue. No doubt, though, there had been government assistance for the new housing.

There was no one around. *They* were all over at the air base, waiting for the evening Christmas show, while *we* were here at their village in -35°F weather, relying on a dubious vehicle to get us back. But we did get back, after another chaotic ride hanging from the straps in Squadron-Leader Watson's snow cruiser.

The evening's Christmas concert was a duplicate of the one at Great Whale, except that the entire audience here was Cree. I asked one of the civilian workers if he could find me a Cree who spoke English. He brought over a cheerful looking young man, and I got my tape recorder ready. I asked him what he did for a living, and he replied that he trapped beaver up and down the Winisk River. When that was out of season, what did he do? Oh, he worked at Toronto airport, driving the little vehicles that hauled luggage around. Though Toronto was nearly a thousand miles south, he spoke as though it were just a commuter ride away. So much for my hopes of a primitive voice from the sub-Arctic for Christmas on CFRB. But I asked him for a "Merry Christmas" in Cree, and he obliged.

The concert over, the officers' bar was an active scene, filled with RCAF blue, the rough-cladding of civilian workers, and the suits and jackets of the news media. A little air force band made some music for dancing with the few women. We heard tales of legendary happenings around the air base, such as how one officer, after a night's cheer in the officers' barroom, staggered outside to retire to his barracks just across the way, fell over the snowbank piled alongside the walkway, and wasn't found until morning, frozen stiff and dead. In their haste to rush to their own bunkhouses, the other chaps had failed to notice him lying there. It was a lesson in trifling with the Winisk weather.

After another attempt at sleeping with the chug-chug of diesel engines going on all night just outside the window, we were up at dawn for the flight home. The ground crew hauled our plane out of the hangar and got its engines going for warmup. It had been an especially cold night, the temperature dropping — to what depth, nobody knew. At breakfast, Squadron-Leader Watson gave a little speech of appreciation for our visit. He looked a bit drawn, but we assumed that was part of the rigours of the previous night's officers' bar celebrations.

On board the plane, relatively quiet while the engines idled, the navigator told us casually about how Squadron-Leader Watson had almost had it the previous night. He'd driven some of the natives home to the village in a panel truck after the festivities and went off the road on the way back, still a mile away from the base. The panel truck stayed stuck in a snowbank. He'd then walked that last mile home. By the time he reached the base, he was covered in frost from head to foot and the next thing to being dead. At breakfast, when he spoke to us, aside from his drawn look, he'd been cheerful and showed no signs of near-death nor frostbite, except for a pale patch here and there on his face. But he'd come perilously close to his last go-round. We marvelled at the toughness of the man and his gung-ho performance to say good-bye to us coming as soon as the next morning.

We took off. We noticed that the North Star just barely cleared the row of stunted little pine trees at the end of the runway. As we soared high, the word was passed around that the ground crew hadn't been able to get the temperature of the engine heads up really to the safe point for takeoff. But take off we did, regardless. It skimmed through our minds what might easily have been a news story of a sub-Arctic RCAF plane crash, with twenty news media types right in the middle of it. But the roar of the North Star's four engines continued, boring through the sky for home. For the next three days after we landed, I could barely hear what anybody was saying. And nobody, including Hutton, said a damn thing about that monumental event: Radio's conquest of the Canadian sub-Arctic.

the apotheosis of
pierre trudeau

I'd accepted the job of Assistant Director (later Director) of Public Information at McMaster University in Hamilton, with a good raise, four weeks' vacation, and regular hours. I was still a journalist, but on the other side of the fence. Even before I was due to leave CFRB, a major news event was about to happen and, still an RB staffer, I was to be there. In April, the Liberal Party of Canada was to hold a leadership convention, the one that eventually made Pierre Elliott Trudeau its leader, and consequently prime minister. The CFRB news team going to Ottawa included Jack Dennett, Betty Kennedy, Bill Hutton, Bob Hesketh, and me, as well as our Ottawa bureau people and a team of our control technicians.

The Ottawa arena site of the convention was jammed not only with Liberal Party delegates, but also with hundreds of news people and observers, not to mention the entourages of the eight leadership candidates. The crowd was equal in size to any NHL hockey game's that might happen there. A sea of faces under funny hats sloped up on both sides of the stands. The high-up bastions of major broadcast media were broadly tagged with big signs spelling out their names. "CFRB NEWS" was prominent.

There was nothing elaborate about the news bastions' seating: just the long, hard wooden benches that came with the old arena. The floor area was crammed with wandering visitors, mostly sporting the red and white colours of the Canadian flag and the Liberal Party, more funny hats, and others just milling about. There was much time to kill yet. There was tension in the air, and the great question of "Who will win?"

The bets were on Pierre Trudeau, Robert Winters, and Paul Martin, Sr., as likely winner. My part in the affair was to cover Winters. The delegates all had their own sections in the seats where most of them sat surrounded by their handlers.

As even the aisles were jammed with people, it was next to impossible to get close to any of them. Fortunately, the Winters group was seated down in front.

I wriggled my way through the jam-packed bodies to stand at the railing just in front of him, dragged down by the big tape recorder slung at my side, squeezed by the delirious crowd, and holding up my mike to catch some cogent Winters remarks, if any, before and after he was chosen leader. I shouted the odd question at him in the din, he gave innocuous answers and remained relaxed and smiling. Farther up the seating was Pierre Trudeau, the minister of justice, with my old University of Western Ontario classmate Bob Stanbury sitting next him. Bob, also a Pearson cabinet minister, had been instrumental as Pearson's liaison with Trudeau in the process of inviting him over to join the Liberals.

After a while, before the voting began, Trudeau magically disappeared. The votes were to be tallied by computer, a first in Canadian elections. A row of IBM cabinets was set up behind the huge curtain at the back of the central stage. A friend of mine from our days at the

Toronto Star, Austin "Ozzie" Winch, another Western Journalism grad, then with IBM, was in charge.

When the first ballot was fed into the machines, the pesky things refused to work properly. This was early in the development of computers as fully reliable bean counters, risky when they weren't set up in their home environment. Ozzie had to ask the convention chairman to announce a delay while his crew put things right.

They were already working at it feverishly to do so.

I managed to wriggle my way across the floor to the computer line. Ozzie was keeping his cool, but obviously, with the entire Liberal Party of Canada waiting to hear whether their man had been picked, he had a lot riding on his shoulders. He gave me a few words on tape while his technical people worked on the computers. Then I wriggled my way back across to the Winters camp.

Hours later, it was announced that two of the eight candidates had been eliminated in the first ballot and the second ballot was under way. The computers were working well by now but the voting process among the delegates was glacially slow. After the third ballot, the victory of Pierre Trudeau was announced. The screams of joy threatened to bring down the arena's steel roof.

Now I understood why Trudeau had disappeared from the stands. He'd been hustled to a back room to be ready for an inevitable victorious re-entry into the arena as the new PM. A brass band blared deafeningly and triumphantly. The huge doors at one end of the arena swung ponderously open. Brilliant light burst forth into the arena, as though the new leader were already deified and was about to descend to Earth. The lights turned out to be the spotlights of two crews of CBC cameramen with their cameras trained on Trudeau from two sides as he walked alone into the middle of the floor, bathed completely in bluish light, an arm raised in victory like a Roman emperor, smiling broadly at his hysterical minions as he paced to the stage.

When he got to the microphone, the accolades went on for ten minutes.

I stayed at the arena long enough to hear what Trudeau had to say, speaking to the delegates unprecedentedly in both English and French, a harbinger of what was to come in his future program of bilingualism, a word as yet not publicly coined. The CFRB operators were taping it, so I didn't have to. In the CFRB hotel-room base, hooked up to the Toronto studios, I did two or three reports on the show from a lowly reporter's view.

When I got back to the CFRB bastion, I was handed a note that said Lester B. Pearson (now former prime minister) would be holding a press conference in the East Block on Parliament Hill. My work at the arena done, leaving the rest up to the more able Dennett, Kennedy and Hesketh, I got a cab and dashed to the date with Pearson. It was to be the prime minister's swan song.

Contrary to what the average voter might think, there was no rush of news people over to the East Block at the behest of the, after all, still *official* prime minister. The main attraction was at the arena. That's where now reigned the new darling of the news, especially in the eyes of the CBC, which over months had literally boosted Trudeau into fame with perpetual love-ins about the new sensation's virtues.

In a dim old auditorium in the East Block, there was only a pathetic turnout of about half a dozen news people, including me.

Pearson emerged from the wings of the stage entirely alone and sat down at a desk with a microphone on it. The man who'd served as president of the United Nations Security Council, who'd brought down the great Dief, who'd created a new flag for Canada, who'd overseen Expo '67 as a world event, who'd originated NATO, who'd thought up UN peacekeeping and won the Nobel Peace Prize for it, who'd later hold forty honorary degrees from universities, who'd drawn huge crowds in past tours, who'd been the prime minister of Canada for over five years was now alone on a dark stage with an audience of six bored reporters — mostly juniors — looking on, no Liberal Party presence visible.

Pearson really had very little to say beyond wishing his successor well and being firm in his belief that Mr. Trudeau would serve his country brilliantly. I didn't even bother to tape those predictable words. Nobody asked a question. The swan song of one of the most colourful prime ministers since John A. Macdonald up to that time was muted, almost inaudible. Pearson stood, gave a short wave — which nobody photographed — and walked off into the wings. The rest of us filed out of that sombre setting, few saying a word.

Meanwhile, back at the arena, the hysterics were continuing. The brass band apparently had never stopped blaring. The CBC's spotlights never left Trudeau. After all, the public network had projected Pierre into the limelight in a way that no one else ever was, and now they were going to keep him there. Pierre was the new boy from Quebec who'd defied Parliamentary custom and worn jazzy shirts and sports jackets and sneakers in the House. Not many months passed before they were on the outs, their affair cooled — over what, I don't remember.

After a few hours of celebration, in which retiring Prime Minister Pearson did not put in an appearance, the word was passed around that the new Prime Minister Trudeau would hold a press conference at the Press Building on Wellington Street, across from the Parliament Buildings. Betty Kennedy was fishing for a private interview with him, and I believe she finally succeeded, despite the splendid isolation from private radio interviews that the Trudeau faction attempted to adopt.

The rest of the CFRB crew was doing whatever they were geared up to do. I shot off to the Trudeau press conference in a cab shared with a bevy of Quebec reporters, who babbled jubilantly in French on what the future held for them in an Ottawa now changed forever by the apotheosis of Pierre Eliot Trudeau. The new millennium had already arrived for them in spades.

The mood at this event was light years away from the Pearson farewell. All was brightly lit in the Press Building conference auditorium. The room was packed, with many standing around the walls. The stage was brilliantly lit. In due course, the PM-elect entered and took a seat on-stage. With him was Mitchell Sharpe, one of the failed leadership candidates, now reconstituted as a staunch Trudeau ally.

As the questions started, I noticed that sitting some distance behind Trudeau and Sharpe was a sombre-looking individual with very dark hair and a massive overhanging forehead. He wasn't called upon to say anything as the press conference proceeded. He didn't seem to comprehend what was going on. I asked the person sitting next to me who that was. She said his name was Jean Chrétien. She didn't know what his role was with Trudeau. As he didn't speak

English, he understandably was in the dark. Some would say that would be proved out decades later.

The whole Ottawa show finally wound down. Some of us news freaks toured hotel rooms, where parties of varying enthusiasm were going on, hoping to pick up some colour for the news. The CFRB crew held its own party in one of its hotel rooms, a tradition among the RB veterans when a major assignment was over. Little by little, as dawn neared, we all faded away to our own rooms.

The Big Party was over.

My final departure from CFRB, of course, meant literally nothing to the continuing story of Canada's most successful radio station. RB and radio would continue on their careers. I would, at forty-two, segue into the more peaceful — or so it was generally believed — world of the University. I was told I was too young to retire. In fact, peaceful retirement was not to be the case in that rapidly expanding political-activist-oriented world of higher education in the 60s and 70s.

But the CFRB crew gave me a great sendoff in the main foyer of the building, with Gord Sinclair reading off a sparkling testimonial to the great work done by Gil Murray in his CFRB career. I was flabbergasted and almost speechless in my stammering reply. Clutching my going-away presents, I fled after a few handshakes. CFRB was free to continue its remarkable role in Canadian radio.

from whence did it come?

S o where or what did this thing called radio spring from? Did
Alexander Graham Bell just accidentally give the wrong phone
number one day and find himself talking to someone thousands
of miles away? Hardly. Many people were involved in bringing about
the phenomenon called radio broadcasting.

Following the same airborne path blazed by the mythical Stentor
and the sky-soaring Daedalus and Icarus, inventors strove to harness
physics so that humans could ride the air and conquer the ancient
handicap of limited human reach. It was millennia from the time of
Stentor's chesty bellows to the exciting early twentieth century days of

Guglielmo Marconi, Nikola Tesla, and Reginald Fessenden — plus anonymous others — when a human voice did finally ride the electronic airwaves.

There was no single inventor of radio. Then who did bring this magical medium into actuality? Three individuals, so far as is known — British, German, and Italian — over a stretch of thirty-five years advanced radio from a nineteenth century theory of electromagnetic waves to the sound of an actual voice that leaped the Atlantic. Scottish physicist Clerk Maxwell, way back in 1864, theorized that electromagnetic waves in space ranged from light — the shortest wavelength — to radio, the longest. Heinrich Hertz, in 1888, produced radio waves by electricity, and found they conformed exactly to Clerk Maxwell's theories.

In 1901, Marconi, who'd already sent radio signals across the English Channel in 1898, was on hand at his receiving station in Newfoundland to pick up a wireless-ed "S" Morse code signal from Britain. Marconi thus proved that not only could wireless radio signals span the distance from Europe to North America, but also that the signals could be "bent" around the earth's circumference. Marconi, Italian but working in England, earlier proved that communication by radio waves was fully possible. Ship-to-shore wireless telegraphy soon linked seagoing vessels with their land-based ports-of-call, using Samuel Morse's dots and dashes.

As early as 1893, the wild genius Nikola Tesla demonstrated the principles of radio signal transmission. Even as Marconi was tinkering with long distance wireless signals, in New York City's Madison Square Garden, Tesla was showing how he could control a model boat in a pool with radio signals. Even so, U.S. Navy observers were at a loss to see how any possible use could be made of such a robotic device in wartime. What a far cry from the 1990s Gulf War and later conflicts, when "smart bombs" hit targets with the guidance of radio and radar!

Discovery of the key to *voiced* radio as we know it came in 1904 with the invention of the thermionic vacuum tube by Sir Ambrose Fleming, of Britain. At last Stentor of legend was being displaced as the real voice of the universe. Lee De Forest, an Iowan, placed a grid in the thermionic tube and gave it better control of electron flow. De Forest's work was a key to early audio and video broadcasting.

So then, when did radio go commercial? It was a Canadian, Reginald Aubrey Fessenden, who invented the system of Amplitude Modulation (AM), the basic radio method that served commercial radio exclusively through to the 1950s. As Frequency Modulation (FM) radio became possible it gradually displaced AM by 2000. Using his AM system, on Christmas Eve 1906 Fessenden broadcast from his station at Brant Rock, Massachusetts, what was undoubtedly the first radio *program* in North America. Over the air, Fessenden read from the Bible, played "O Holy Night" on a violin, and followed with Handel's "Largo" on a phonograph played into his crude microphone.

Even though his listeners were mainly sailors on United Fruit Company ships equipped with primitive voice wireless ship-to-shore receivers, Fessenden lifted what was a dry communications medium to one of actual entertainment, the hallmark of the radio era to come. The sailors were probably astounded by what they heard. Fessenden, incidentally, also invented microphotography, and a radio pager set into the hardhats of Brant Rock technical staff.

Fessenden was a Canadian child genius. Born in East Bolton, Quebec, in 1865, he'd read *The Rise And Fall of the Roman Empire* by the time he was seven. At sixteen, he was *teaching* Greek and French at Bishop's College School in Lennoxville, Quebec. In 1886, he met Thomas Edison, and became Edison's chief chemist in the inventor's New Jersey labs. There the versatile Canadian genius invented a fireproof rubber insulation for electric wire, and even made the surfaces of varnishes harder and shinier. He also built a special dynamo for Edison's motion picture equipment.

After leaving Edison, Fessenden went on to invent a new kind of light bulb for George Westinghouse, circumventing Edison's patent, and a new kind of steel for motors, using silicon instead of carbon; and he was named to a chair in electrical engineering at Purdue University. By intensive research into radio signals, he eventually patented five hundred inventions involving radio.

Amplitude Modulation was his greatest achievement. All commercial radio stations around the world, from the early 1920s to the 1960s, were broadcasting on AM. Strangely, the name and accomplishments of Canadian Reginald A. Fessenden remain little-known and little-honoured publicly in the pantheon of inventors. Accolades usually go to U.S. and British inventors, but it did take a Canadian, in a time when Canadians were universally typed as hewers of wood and drawers of water, to ride the waves of the future on the phenomenon called radio. Still, accolades for Fessenden seldom come.

Another Canadian technical genius, Edward S. Rogers, CFRB's founder, pushed the evolution of radio well ahead with his alternating-current power vacuum tube. In 1912, when he was aged eleven, Ted had his own half-kilowatt spark transmitter, one of the first licensed amateur sets in Canada — even in North America. At twenty-one, he was the first Canadian to send a radio signal across the Atlantic. In 1925, with his AC power tube, in one stroke he not only eliminated the dangerous battery power source, but also abolished the immobility of battery-powered radio sets.

Rogers' invention opened the way not only for more convenient and simpler home radio consoles, but for small portables as well. Elimination of heavy batteries allowed lighter-weight radio sets in airplanes, more easily-carried units for military field use, and early military computers — a boon in the coming Second World War — ultimately leading, as much more time passed, to the concept of lightweight radio-controlled systems for spacecraft, and the miniaturizing of radios in general. Only the arrival of the transistor in the late 1950s eclipsed the radio tube as the heart of radio transmission.

One might have thought that Ted Rogers would have received wild international acclaim as a new young Edison. One would have been wrong. At first, North American radio manufacturers didn't buy in to this young Canadian's argument that his invention would remove householders' aversion to the existing battery radio sets and ultimately put one into every home. Rogers maintained that, once rid of the wet battery, with its tendency to explode, and instead using the home's system of alternating current, safer than direct current, the radio receiver would be as widely accepted as the telephone. But, of course, adopting the Rogers power tube meant a sales delay while circuits were rewired in the products already being sold. That would slow sales and cut profits. Rogers got nowhere.

Being Ted Rogers, however, he didn't stop at "nowhere." Along with his father and brother, Elsworth, he bought rights from the

Independent Telephone Company of Toronto to build radio sets. He then designed a radio wired for the AC power tube and displayed it at the Canadian National Exhibition in Toronto in August, 1925. The "Rogers Batteryless," later the Rogers Majestic, quickly sold.

The Rogers' turned out hundreds of sets, then thousands. They sold easily in both Canada and the United States. As Rogers had long before predicted, sales of Batteryless radios zoomed. Rivals who'd rejected the Rogers invention had to redesign their new lines of radios. It was an electronic revolution. But Ted Rogers' name is not enshrined as a hero of international radio history.

And what of the living room boxes that sat in a corner and made all those sounds? The radio receiver had a remarkably short evolution from a tabletop junk pile into a desirable item of fine furniture. Once the news dawned on manufacturers that sets controlled with knobs and dials, and not "ticklers," were what the public wanted, the race was on to sell the most radios to the most people. As with many raw inventions, such as the automobile with its early crude controls and lack of springs, the time comes when a manufacturer must pay closer attention to female tastes over the rough-and-ready acceptances of males. Style and ease-of-use come to mean almost as much as performance. So, within only a year or two of radio's debut as a viable entertainer, the junk pile was housed in fine wood cabinets specially designed to be identified exclusively with their makers.

Some of this psychology was due to the advancement of radio theory by people with training in science, and the advantage of the pioneering work that had gone before them. The crystal set's tickler and crystal gave way to a system that could manipulate an induction coil to "tune" the oscillations of a particular broadcast frequency. Now a specific radio station could be brought in by just turning a knob on the front panel of the set. In the earliest radios, familiar in old photos of the tabletop Atwater-Kent, an oblong box with a bakelite front panel displayed a row of wide knobs, or "attenuators," their edges marked off with graduation lines.

One knob was for locating a station's frequency. The others were necessary to achieve the finest tuning-in of a station possible. Some controlled "detectors" that caused electrical current to flow in one direction only, allowing pickup of the low frequencies generated by microphones of the time. It took some skill for the listener to balance the attenuators so that they combined to bring in the clearest signal.

This, of course, was too technically bothersome for the casual listener. Further development brought a "ganging" of the many attenuators, and a reduction of the knobs to two or three, one for tuning, one for "sharp" or "low" sound, another for additional fine tuning. These all, of course, eventually were boiled down to just one tuning knob and a volume control.

Next came the "super-heterodyne" receiver, simplifying the selection of wavelengths and reducing static and other interference. Automatic volume control was achieved by introducing a circuit and a diode tube that leveled off waxing and waning radio signals, producing a reasonably constant reception level.

As time passed, the "peanut tube" and smaller speakers with greater audio range appeared, making the small portable radio possible and popular. Then, in 1940, came the pushbutton system that saved the bother of dialing. Beyond even this, in the 1970s, the tuning knob itself was banished. Pushbutton tuning and volume control became standard, mostly thanks to the transistor's appearance in the 1950s and the microchip's in the 1970s. Finally, remote control banished even the effort of leaving one's chair to do anything.

When dressing up this increasing jumble of radio works in beautiful cabinets came into vogue in the mid-1920s, radio sales shot up. Not only had the old battery-dependent radio sets been abolished by the Ted Rogers power vacuum tube, but the very idea of a radio as a decorative furniture piece gave the old tinkerer's box a new lease on life with the public. The radio as fine furniture in oak, mahogany, and Australian mulga wood cabinets was something a family could be proud to show off. Before that, some people were even putting their radios away for the summer!

Besides the Rogers Majestic, one of the first six-tube radios, there were De Forest Crossley, Gold Medal Radio-Phonographs, Northern Electric, Marconi, Atwater-Kent, Stewart-Warner, and other popular makes. All of them produced stand-alone consoles, growing turned wooden legs and displaying polished surfaces like the best traditional furniture classics.

In the 1930s, the legs were eliminated in favour of the more "modern" style enclosed down to the floor in sleek inlaid wood cabinets of "streamlined" design — that is, with no square corners, like the new mid-30s automobile models. The floor-length wooden skirt of the console radio was actually necessary to allow for and conceal a much larger loudspeaker. The single speaker was now widening to twelve and fifteen inches to produce more bass sound, even if in a more distorted form in lower ranges. The thinking was that music sounded richer through a "boomier" speaker. By the 1950s, further evolution catering to public taste shrank the home radio back to about the size it had been in the 1920s. This literally "put the radio on the shelf" in the postwar smaller houses and apartments. Radio came out of its big fancy box again, and the era of the fine furniture piece was over.

Critics of the older sound systems point out that the audio quality of the console radios was low mainly because of the cost of having the cabinets made by outside firms, plus profit-taking along the way from manufacturers through jobbers and wholesalers to the final retail level. When it was all tallied up, radio components themselves had to be made at low cost — with resulting lower sound quality — for production and sales to be feasible.

Long ago now, the costs of the many stages of marketing of most kinds of goods has been routinely analyzed, refined, and squeezed for maximum return. Actually, the public ear didn't demand better radio sound. Most were happy with the "boomy" 1930s radios, regarded as a relief from the scratchy, screechy noise of the 1920s sets, the mechanical phonographs, and their records. To lower the prices of radios to levels the Depression-era public could afford, "frills," if not labour as well, were cut so that the once generous features of the product became marketed as "extras," or totally dispensed with — usually the latter.

That's what happened to the radio console cabinet, and also, incidentally, to the early television set cabinets. In radios, as public appetite grew for better-quality sound in the new era of hi-fidelity stereo FM, those perks had to go. In the early 50s, the first television set I owned was part of a "home theatre" with a sixteen-inch TV screen in a walnut-veneer cabinet the size of a small refrigerator laid on its side. Also in the cabinet was an AM radio, an RCA Victor 45 rpm record player, a 78 and 33 1/3 rpm record player, and a twelve-inch speaker. A few years later at the same store, you could buy only a twenty-three- or twenty-eight-inch TV in a box just big enough to hold it.

The onset of FM radio and "hi-fi" stereo, generated by the LP and its high-quality microgroove audio reproduction in the 1950s, brought in the idea of setting two stereo loudspeakers apart at some distance from each other and housing the radio components in long, large, coffin-like cabinets. "Hi-fi nuts" wanted sound quality almost beyond the capability of the human ear.

Disparagers remarked that only dogs could hear the best of the system, giving new meaning to the old RCA Victor trademark of the white mutt with ears cocked to "His Master's Voice." Specialty stereo manufacturers then began to produce hi-fi components that could be plugged together, the hi-fiers often mixing the products of different stereo makers to create the system that they felt satisfied their acute hearing. "Woofers" and "tweeters," the hi-fi nametags for bass and treble loudspeakers, became household words.

Record players turned into disc platforms, sitting on bases often filled with concrete to give rock stability. Sometimes a bedroom closet would be turned into a giant speaker cabinet with the speakers set into the door, and the walls of the closet coated with thin concrete for maximum sound bounce through a port cut farther down in the door. Rooms were stripped of drapes, curtains, rugs, and soft furniture so higher audio ranges would not be soaked up and made inaudible to keen ears.

The Rogers power vacuum tube revolutionized the radio business in the 1920s and 1930s, but a giant step was yet to come that would equal and surpass that achievement in scope and even in social impact. The invention of the electronic transistor in 1947 turned the world of electronics on its ear. This tiny component that could be held on a fingertip bore a power that instantly made obsolete the inner works of all vacuum-tube radios, computers, and anything electronic. It would cut down some devices to the size of cigarette packages, eliminate the old problem of heat burnout of circuit wiring, and win the Nobel Prize for its inventors. Circuits printed in copper on boards added to the revolution.

Where glass-enclosed vacuum tube grids for decades had controlled the flow of electrons in conventional devices, the "point-contact resistor," as the transistor was called in its first version, projected the world in one stroke into a new era: the age of the miniaturized computer, the miniaturized radio, and the longer-lasting TV set, even launching the space age. Radio fans could carry with them a set whose capacity would have required a floor console to satisfy them in pre-transistor times. To many, a "transistor" is just a small portable boom box that can make a big noise and be the bane of riders of public transit.

By abolishing the vacuum tube, the transistor — in its second version called the "junction transistor" — at once solved the problem of size and heat, two of the previous most limiting features of electronics. Computers such as the ENIAC and the Second World War's Colossus — which had weighed thirty metric tonnes and taken up two thousand square feet (one hundred and eighty square metres) of floor space — could be reduced to desk-top size, then later to vest pocket dimensions, and later still, in many specialized functions, to little more than a pinhead.

The transistor was tiny, but its implications were huge. It became possible to build computing devices and radio relays small and light enough to make spacecraft and space travel feasible. For radio, the potential was immense. No longer would an electronic device burn out early from the heat of vacuum tubes. That prolonged its life enormously. Radio stations could operate around the clock and still stay cool.

The trio of Bell Laboratory researchers who changed electronic history, William Shockley, John Bardeen, and Walter Brattain, used a germanium rectifier having metal contacts and a needle that touched a crystal, to be used as a semiconductor. A month after this first transistor version was announced, Shockley produced the junction transistor, which transferred current across a resistor with a silicon microchip. They jointly received the Nobel Prize for Physics in 1956. Bardeen became the first man to win the Nobel Prize for Physics twice.

war on the airwaves

Aside effect of radio — under the British term "wireless" — was its use in the military, where, after its introduction for field communication in the First World War, it developed rapidly into an essential weapon for all of the world's armed forces. In the 1914–1918 conflict, it was almost exclusively telegraph-key operated. Though the "bug" went on being used even through the Second World War in a more updated form, voice radio came into military use between the wars.

This was, however, largely limited to relatively short distances. Atmospheric conditions, with their static (called by military signalmen

QRM) and fading so often impaired voice transmissions, garbling message content, that the short-wave wireless key telegraph system was found to be more reliable for long distance military communication, extending even to international diplomatic messaging.

It also had the advantage of needing a trained wireless operator to receive messages. While these weren't scarce, the chances of an untrained enemy intercepting an important voice message were reduced. Military radio also gave birth to codes designed to thwart enemy eavesdropping. In some cases, codes did just this. But beavering away just as diligently, cryptanalysts on both sides worked to break enemy codes as fast as new ones were created. Such was the case in the Second World War, both in the war in Europe and the war in the Pacific.

Allied eavesdroppers, including Canadians, constantly monitored the military communications of the German and Japanese military forces so successfully that, postwar, top commanders of all sides credited these operations with playing a crucial, even a *decisive*, part in the Allied victories.[2] Members of prewar Canadian commercial radio, converted into soldiers, sailors, and airmen, performed military roles as the war went on. Your author was one of them.

Having logged considerable time in radio before 1944, when my time came to join the army, I eventually found myself a member of Number One Canadian Special Wireless Group of the Royal Canadian Signals Corps, serving in the South West Pacific theatre of the war, together with three hundred and thirty others, as an eavesdropper on Japanese wireless communications. In fact, Allied operations in the Pacific war were far more dependent upon monitoring and cracking Japanese coded messages than in the European theatre. Planting Western spies in Japan and its conquered areas was next to impossible, due to the recognition factor of whites in the Japanese population.

In Europe, it was possible to insert Allied spies into enemy territory who could report back on German military operations, even by wireless, with a good chance of escaping notice. In the Pacific, all Allied operations depended upon intercepted wireless reports of Japanese troop and ship movements picked up by diligent wireless interception

---

2   See *The Invisible War* by Gil Murray, Dundurn Press, 2001.

of enemy signals. It was the only way to find out what the enemy was up to, and it proved devastatingly successful for the Allies.

Codes devised by the Japanese before and during the war for sending their vital orders to their far-flung wartime units throughout the Pacific theatre were broken and already known to Allied signals intelligence units. Unknown to the Japanese, Allied wireless operators and decoders were copying and reading their thousands of daily military messages and relaying them to Allied battle units for action, which was decisively taken. Japanese ships were sunk by the hundreds, and troops were devastated by Allied air attacks.

The top Japanese general in charge of troop communications stated postwar that he and his staff, confident in the safety of their codes, were totally unaware that Allied commanders were acting on their decoded messages. In the Pacific, at least, radio in military form was the nemesis of Japanese imperial expansion, acknowledged later by all leaders as a key weapon in winning the war.

the sound of money
in her voice[3]

I f radio were female, the above words would be quite appropriate to describe what made radio work. Without advertising's very large money source, private commercial broadcasting — expensive even from the beginning years — could not have come about. Commercial radio and advertising, or government sponsorship, go hand in hand, horse and carriage. You can't have one without the other. They are mutually dependent, born about the same time, and growing up into mighty money machines.

There was yet another actor lurking in the wings, who came forth early and joined the others in making it all work: the phonograph

---

3   See *The Great Gatsby*, by F. Scott Fitzgerald.

record. The billions of miles of grooves full of music that have since saturated the once-silent airwaves gave radio its substance for over half a century, until, on some stations, talk-radio mercilessly blasted the sound of music out of its way. Still, it almost goes without saying that music, whether live or recorded, continues to be the mainstay of radio's full-programmed broadcasting content apart from talk-radio's loquacity.

Was the phonograph record the child of the radio–advertising marriage, or was it private radio's proud parent? Just as advertising earned the money to sustain radio, the phonograph record was born of the same physics — electronics — that enabled radio to materialize. From the start, it was obvious that radio and the record needed each other. While the phonograph could provide home entertainment without the help of radio, what lay in those bumpy grooves became known and popular mainly through hit records played on radio. The two media then grew up together.

From Thomas Edison's crude tinfoil-on-revolving-drum, through cylinders, to Emile Berliner's thick flat disc, to Eldridge Johnson's thin composition 78 rpm record, and on through the 33 1/3 LP to tape and CD, the continuously evolving recording industry helped to orchestrate the tumult of twentieth century change. Radio and its advertising mate evolved too, but in more subtle ways. Together with their cousins, moving pictures and television, they so profoundly changed the world sociologically that there is little similarity between the start of the twentieth century and its end.

Until about 1950, the public wasn't particularly concerned about the relatively mediocre sound quality of radio and records. As only the concert-going population knew what "live music" from an orchestra or vocalist really sounded like, the vast majority was content with primitive recorded and broadcast audio. Television, in a roundabout way, was going to force a major change in that.

To compete with the little screen that was soaking up large parts of its audience in the 1950s, the film industry introduced wide-screen movies along with extremely high fidelity sound recorded on film more stringently than ever before. The effect on audiences of the big, wide screen and the loud, realistic hi-fi soundtrack was overwhelming. People began drifting back to the movies partly for the stereo sound.

Audio engineers often had said they could produce much better sound than what emanated in the 1940s from AM radios and electric record players by redesign of electronic circuits and taking more care in making components and connections. High-quality sound was already there in the record grooves, they maintained. It was the mediocre sound-reproducing ability of home record players that was at fault. Manufacturers, however, were happy with the steady sales of their low-fi sets and were slow to make changes. They had to, however, when the LP and FM radio came along.

Although an early stab at bringing in stereo recording was made in the U.K. in 1933 with a double-grooved 78 record, it never caught on with the public, probably because of the mono and low-fi of AM radio. An attempt by RCA Victor in about 1940 to sell a phonograph minus a needle, using a beam of light, failed. The idea of a light beam riding the record grooves was valid, but the particles of dust in the air of most homes spoiled the sound quality. Static caused by the minute particles floating in the beam between the pickup and the record grooves made the music unlistenable.

Even though they'd gone into production of the light-beam phonographs, RCA scrapped the idea. Not until the mid-1950s did things pick up, when the Audio Fidelity company in the U.S., simultaneously with the Pye and Decca companies in Britain, produced a better-sounding recording. This coincided with the emergence of increased FM broadcasting outlets, usually as "little brothers" of the established AM station, operated from the same premises.

In a minor revolutionary radio experiment, for awhile in the mid-1960s it was possible to hear a broadcast in actual stereo at home by tuning in to CFRB on an AM radio and simultaneously to CFRB-FM on an FM set. This was eventually replaced by a fully stereo Standard Broadcasting FM stereo station, CKFM, but for a while listeners could enjoy the novelty of hearing on-air music in stereo before FM radio became common.

As noted earlier, Toronto's CHFI was among the first, if not *the* first, of all-FM radio stations in Canada. Soon, almost every AM station had its FM shadow. By the 1990s, radio was phasing over into FM entirely and gradually out of AM.

Among the converts was CBC Radio, which shut down its venerable AM station, CBL Toronto, and converted to CBC Radio One and Radio Two, both FM.

Radio has done a remarkable job of surviving for more than a hundred years, from its first feeble sparking signal produced by Marconi, to the twenty-first century's super hi-fi stereo — probably truer and of wider range than most human ears can detect. It has not only survived, but has steadily improved its quality and range, kept its commercial viability in the face of competing media, and become an even stronger force than its competitors because of its ubiquity and mobility.

It is everywhere. It travels inside the maze of modern automobile traffic; it walks within the earphones of the young and old on the streets — in mysterious outer silence, but inner joy; it is displacing the wired pathways of endless telephone talk with the wireless cellular phone; it lets man on earth talk with man in space, or gives orders from the ground to manipulate unmanned spacecraft and airborne "smart bombs." Predictions that television would render radio obsolete not only drastically failed fulfillment, but spurred radio on to equal its rival in audience share.

While film and video leave nothing to the imagination, it is imagination itself that made radio the friend of every person. Before television, radio plays evoked pictures from the mind, blindly stimulating human imagination as only books could do, causing listeners to create some of the play's effects for themselves. Even the voices heard on the air tweak musings on what their owners really look like. The demise of the radio play diminished that exercise of the mind.

If any cultural force reshaped the reality of living, it was radio. Refinements such as television and cellphones intensified this reality. As one of the most significant of human achievements, the first century of radio should surely have been celebrated on a world scale.

I don't think it was.